2008

*Cleveland's Transit Vehicles*

# *Cleveland's*
EQUIPMENT AND
# *Transit*
TECHNOLOGY
# *Vehicles*

JAMES A. TOMAN &

BLAINE S. HAYS

The

Kent State

University

Press

*Kent, Ohio,*

*and*

*London, England*

© 1996 by The Kent State University Press,
Kent, Ohio 44242
ALL RIGHTS RESERVED
Library of Congress Catalog Card Number 96-11223
ISBN 0-87338-548-9
Manufactured in the United States of America

02 01 00 99 98 97 96   5 4 3 2 1

*Frontis:* In the greatest hour of public transportation in Cleveland, a bronze Moses Cleaveland watches over Public Square. Behind the statue, car 4118, from the city's most remembered fleet of streetcars, waits for passengers to board. The Hotel Cleveland forms the backdrop for this 1935 scene. *(Cleveland Railway photo, Blaine Hays Collection)*

*Library of Congress Cataloging-in-Publication Data*
Toman, Jim.
Cleveland's transit vehicles : equipment and technology /
James A. Toman and Blaine S. Hays.
p.   cm.
ISBN 0-87338-548-9 (cloth : alk. paper) ∞
1. Street-railroads—Ohio—Cleveland—Rolling stock.
2. Trolley buses—Ohio—Cleveland.  3. Buses—Ohio—
Cleveland.  I. Hays, Blaine S.  II. Title.
TF920.T64  1996
625.6'6'0977132—dc20           96-11223

British Library Cataloging-in-Publication data are available.

*To our mothers:*

*Lillian W. Toman (1903–1993)*

*&*

*Fern D. Hays (1921–1993)*

# Contents

Acknowledgments ix

Introduction xi

Abbreviations Used in the Roster xiii

CHAPTER ONE

**Technology in Cleveland Public Transportation**

*by J. William Vigrass*

3

CHAPTER TWO

**Roster of Cleveland's Transit Vehicles**

21

PART A: *Cleveland Streetcar Roster from the Time of Electrification* 23

*Section 1:* Outline of Electric Streetcars Owned by Predecessors of Cleveland Railway 24

*Section 2:* CER and CCR after the Consolidation of 1903 26

*Section 3:* Cleveland Railway—Passenger Equipment Inherited from Predecessor Companies 28

*Section 4:* Passenger Equipment Purchased by Cleveland Railway and Cleveland Transit   57

*Section 5:* Trailer Cars   108

*Section 6:* Service Cars   127

PART B: *Cleveland Rapid Transit Roster of Equipment*   163

*Section 1:* Shaker Rapid Transit   163

*Section 2:* Cleveland Rapid Transit   178

*Section 3:* Rapid Transit Service Cars   189

PART C: *Roster of Trackless Trolleys*   193

PART D: *Roster of Cleveland Motor Coaches*   206

*Section 1:* Buses Purchased by Cleveland Railway Company   208

*Section 2:* Buses Purchased New by the Cleveland Transit System   225

*Section 3:* Buses Purchased by Cleveland Transit from Other Transit Agencies and Motor Coach Service Units   250

*Section 4:* Buses of Regional Transit Authority Acquired from Other Operators or Purchased New   254

CHAPTER THREE

**The Detroit-Superior Bridge Subway**

*by Jack Ainsley*

267

# *Acknowledgments*

This book stands as a companion to the chronological history of Greater Cleveland transportation presented in the first volume, *Horse Trails to Regional Rails: The Story of Public Transit in Greater Cleveland*. As such, the authors are grateful to and acknowledge the contributions of all those listed there. Others, though, have made a particular contribution to this volume, and we acknowledge them here.

First of all, thanks to friends and fellow photographers and videographers Steve Heister, Tim Krogg, Chuck Bode, Bruce Dzeda, Bill Herrmann, Jeff Marker, and Larry Podwoski.

We also wish to thank those who made their photographic collections available to us as we attempted to illustrate this roster of Greater Cleveland transit equipment.

In particular, we thank Herb Harwood, Ralph Pfingsten, Jim Spangler, Bill Vigrass, and Bruce Young. Thanks also go to several members of the CTS/RTA team who provided us with photo resources: Nick Barone, Barbara Branch, Lew Childress, Sal Felice, Grace Keener, Corman Moore, John Seman, and Charles White. Thanks also to Tim Lassan for sharing photos and artifacts. And we remain especially indebted for the work of Doc Rollins and Ross Barnard, official photographers for the Cleveland Railway and Cleveland Transit System.

Several other people played an important role in helping with the text of this volume. We wish especially to thank Bill Vigrass, who volunteered to write the section on transportation technology and who also contributed material to the roster. We are very grateful to Bob Korach for his invaluable and painstaking help in providing data for the roster, and then for his care in proofreading and critiquing it. For latter-day roster details, we sincerely thank Robert Humanchuk who gave a hand at a critical time in roster development. Thanks also to Robert Landgraf and Dale Hohler. We also acknowledge Jack Ainsley, who back in 1965 recorded his recollections of a ride through the subway of Cleveland's Detroit-Superior Bridge.

We are grateful to various RTA staffers who helped us collect our data: Darlene Boldware, Cal Cross, Ted Donahue, Tom Fessler, Jack Gilchrist, Linda Green, Gary Grugel, Dan Hurley, David James, Wilson Jerdine, Mila Hyvnar, Chuck Kocour, Mike McKenna, Tim O'Donnell, Ed Opett, Rhonda Raidl, Boyd Rolle, Tony Russo, Alan Sasala, Doug Seger, Rodger Sillars, Joyce Snyder, Dave Thomason, Al Uhl, Les Varga, Brian Willse, Don Wood, Jim Wysocki, and Ben Zywiec. We also thank our editor at The Kent State University Press, Joanna Hildebrand. Organizing the many pieces of this compendium was no easy task, but she did it with precision.

And again, heartfelt thanks to Joan Hays, who understands just how demanding a taskmaster the streetcar can be for those who are true "juice fans."

Like its companion, this book is the result of real collaboration. The richness of detail is due to the care and commitment of our associates. We take full credit for its faults.

# *Introduction*

Greater Cleveland has legitimately laid claim to many achievements. Not the least among them was its public transit system. From the beginnings of the horsecar era in 1859 to the "sardine days" of World War II, Cleveland transit operators provided high quality service to their patrons while introducing procedures and equipment that were widely copied elsewhere.

The social and political aspects of Cleveland's public transportation history are the subject of this book's companion volume. Our focus here is on the more technological aspects of how the system ran and on the vehicles it used and is using to carry its passengers and maintain its infrastructure.

From the start of street railway operations in 1859 until the end of the surface electric era when the trackless trolleys stopped running in 1963, the city was crisscrossed with hundreds of miles of track and overhead wire, and with thousands of poles to keep the overhead in place. Thousands of streetcars, and then thousands of buses, were bought, and they carried billions of passengers. The old Cleveland Transit System alone carried over 493 million passengers in 1946, and that total does not reflect what the various suburban carriers could have added to the toll.

The glory days of public transit are now in the past. Today only about 60 million riders make use of the county-wide services of the Greater Cleveland Regional Transit Authority. The streetcars and trackless trolleys are gone, but hundreds of buses and rapid transit cars still carry out their vital mandate to transport people who either depend on public transit or who prefer it to the use of their own private automobiles.

Today's natural-gas-powered buses and powerful rapid transit cars are a technological world away from the vehicles which plied the streets of Cleveland when the Tayler Grant gave to the Cleveland Railway Company control over the city's transit operations. In this volume, we describe and list both those early vehicles and the modern ones, and we record the relentless pattern of change from the early rail lines, to the rapid transit, and to rubber-tire operations.

It is not so much a nostalgic look at earlier times as it is a description of how ongoing developments in the industry changed the way the public transportation system carried out its mission. In Cleveland it was carried out with rare efficiency and with good speed.

We confess, however, to one nostalgic piece. Back in 1965, Jack Ainsley, a Cleveland railfan, wrote for one of the authors his reminiscences about taking the streetcar through the Detroit-Superior Bridge subway. We include his narrative in Chapter 3. His recollections help frame the technological focus of this volume and give it a more human face.

# Abbreviations Used in the Roster

| | |
|---|---|
| AC | air conditioning |
| ACC | accident |
| ACQ | acquire |
| AR | arch roof |
| BLDR | builder |
| BLT | built |
| C&Y | Cleveland & Youngstown |
| CCR | Cleveland City Railway |
| CE | center entrance |
| CER | Cleveland Electric Railway |
| CIRR | Cleveland Interurban Railroad |
| CLC | Cleveland-Lorain Highway Coach |
| CON | converted |
| CR | Cleveland Railway Company, 1910–42 |
| CRT | Cleveland Railway or Cleveland Transit type |
| CTL | controller |
| CTS | Cleveland Transit System, 1942–75 |
| CUT | Cleveland Union Terminal |
| CYL | cylinder |
| D | delivery |
| DE | double-ended |
| DR | deck roof |
| DT | double truck |
| F | "fleet," referring to buses, trackless trolleys, and CTS rapid transit cars |
| FEN | fender |
| FN | former number |

| | |
|---|---|
| G | painted green and gray |
| GE | General Electric |
| GM | General Motors |
| HDL | headlight |
| LEN | length |
| LRC | Louisville Railway Company |
| LRV | light rail vehicle |
| MCB | Master Car Builders |
| MOT | motors |
| MT | Municipal Traction |
| MU | multiple unit |
| NORM | Northern Ohio Railway Museum |
| OM | one-man operation |
| OOS | out of service |
| PCC | Presidents' Conference Committee type car |
| PTD | painted |
| REB | rebuilt |
| REN | renumbered |
| RR | railroad roof |
| RTA | Greater Cleveland Regional Transit Authority, 1975–present |
| SHRT | Shaker Heights Rapid Transit, 1944–75 (earlier the Cleveland Interurban and Cleveland and Youngstown) |
| SCR | scrapped (all streetcars scrapped at Harvard Yards unless specified) |
| SE | single-ended |
| ST | single truck |
| T | type, referring to streetcar fleets |
| TRC | Transit Research Corporation |
| TRK(S) | truck(s) |
| TT | trackless trolley |
| TTC | Toronto Transit Commission |
| VD&D | Van Dorn and Dutton Street railway supply (East 79th Street) |
| W | weight |
| WH | Westinghouse |
| ★ | denotes an unnumbered vehicle |

*Cleveland's Transit Vehicles*

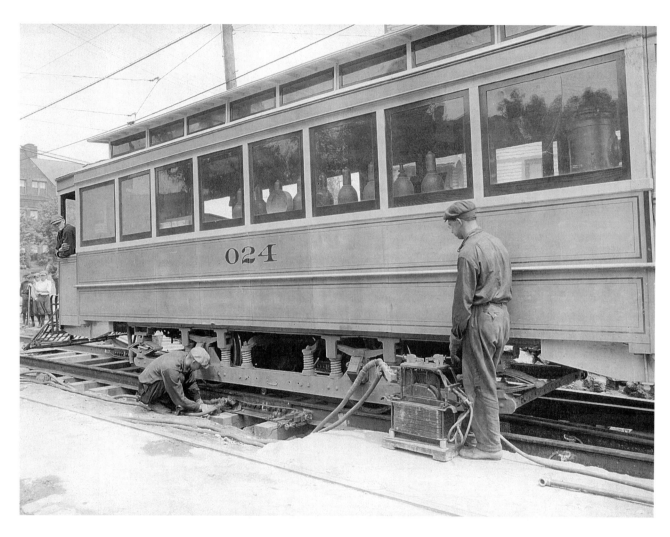

The most common number in Cleveland transit history was 024, appearing on no fewer than five vehicles over time. This 024 was a single-truck welder, seen here on a rail repair call. *(Cleveland Railway photo, Blaine Hays Collection)*

CHAPTER ONE

# *Technology in Cleveland Public Transportation*

J. WILLIAM VIGRASS

### *The Formative Years*

In the earliest days of street railway development, Cleveland, like many cities, sampled a number of technologies. Insofar as literature exists, Cleveland's horsecar lines followed standard technology of the mid–nineteenth century. During the latter part of the century, two cable car lines and an early experimental electrical installation of the Bentley-Knight underground conduit system were built.

The beginning of successful electric traction in Richmond, Virginia, in 1886 was followed by very rapid development of technology. Changes came quickly in both car equipment and in fixed wayside equipment.

In the early days, steam-operated power plants generated traction voltage, 500 to 600 volts DC, directly, using reciprocating steam engines to drive dynamos or generators. Because of voltage drop, caused by electrical resistance, the low voltage DC could be transmitted only a few miles. An early expedient was the "booster" at the power plant. A 600-volt DC motor drove a generator that produced a higher voltage, say 750 volts. This voltage was carried by feeder cable to the outer end of a car line, by which location voltage had dropped to 600 volts or less because of resistance of the cable.

The author wishes to express his thanks to Robert S. Korach and Robert J. Landgraf for editorial review, corrections, comments, and additions.

Wet cell batteries were used to meet peak-hour needs. They were charged in off-peak hours when demand was lower and during the night when demand was nil. These power systems were typically owned by the street railway. One of Cleveland's steam powerhouses still stands in the Flats, having been reincarnated as an entertainment center. Its fate was uncommon. Elsewhere, most powerhouses were demolished.

In the first decade of the twentieth century, electrical technology changed dramatically. The substation was developed, first manually and then later automatically controlled. A substation accepted high-voltage alternating current, typically at 13,000 volts or more, stepped it down to a lower voltage, say 460, and then converted it to 600 volts DC, using a motor generator in the early days and later a rotary converter. The latter combined the windings of a motor and a generator on one armature. By the teens, the remote automatic rotary converter had become the industry's standard. It was available in different ratings, but the Cleveland Railway Company standardized on the Westinghouse 1,500 kilowatt (kw) unit. Where more power was needed, multiples of 1,500 kw units were installed.

Rotary converters were used by Cleveland Railway (CR) and its successor, the Cleveland Transit System (CTS). Some remained in use on the original CTS rapid transit after the street railway lines were abandoned. Their use reduced the need for new capital expenditures when the rapid transit was built in 1953–55. In particular, the noise of converters at Windermere and at Coutant Street (West 118th Street) was evident to passengers at each terminal. Others along the way were on streets nearby and were connected by feeder cables to the Rapid. They were replaced by silicon rectifier substations coincident with the airport extension in 1966–68. Having earned their keep, the rotary converters were retired.

Car-borne equipment evolved rapidly from the early four-wheeled cars that closely resembled horsecars. Indeed, horsecars were later often used as trailers, but their lightweight construction did not hold up under the more rapid electric service, and they were replaced with purpose-built cars by the turn of the century.

Early electric cars had simple rheostatic control that soon gave way to series-parallel control with some resistance. Electrical losses were fewer.

Earliest motors used stacks of copper sheets as brushes to convey current to a motor's commutator. They were bent in the direction of rotation and would burr when the motor was reversed. Carbon brushes were soon substituted; these were self-lubricating and bidirectional. Great strides were made in formulation of carbon for specific needs. As of 1990–94, most rail transit and diesel-electric railroad locomotives use DC motors of a nominal 600–750 volts with carbon brushes. Their principles date to the turn-of-the-century's streetcar technol-

This rotary converter at the Cedar Avenue substation at Ashland Road converted 1,300 volts of alternating current to 600 volts DC. *(Cleveland Railway photo, Blaine Hays Collection)*

ogy with steady evolutionary development. All RTA passenger-carrying light or heavy rail cars use DC motors with carbon brushes.

Early streetcars employed handbrakes operated by the motorman. They were a simple development of the horsecar brake: a crank-type handle acting as a spindle had a chain that wound upon it. The chain was connected to brake beams that applied brake shoes to the wheels. They were simple and, at the 4–6 mph horse-drawn speeds, effective enough.

As speeds and loads increased, power braking was needed. Air brakes had been applied to railroad use, so it was natural to apply them to streetcars. The electrical technology of the late nineteenth century did not provide small compact motors to drive on-board air compressors, so stationary compressors at carbarns or terminals were used to charge large reservoirs on cars. Motormen had to be careful not to use all the air before the next compressor station was reached. Handbrakes were still provided as backup. This method was not satisfactory, and by the turn of the century compact motors of 5–10 hp running off the 600-volt traction voltage had been developed, and car-borne air compressors became standard by 1914. In some places they were required by public regulations if the street railway company had proved reluctant. Such was not the case in Cleveland, where the company was a technological leader.

A myriad of auxiliary devices was developed: headlights, Nichols-Lintern tail lights that turned from green to red as power was shut off to the brake, trolley poles, seats, trucks, interior lighting, door control systems, fenders, ventilating devices, couplers, bearings, windows,

Car 1184, equipped with an Eclipse fender, makes its way along the side-of-the-road right-of-way on Clifton Boulevard. *(Bill Vigrass photo)*

flooring, and heating were all developed in a decade and refined in another decade into reliable systems. Progress was rapid and came to be expected by the public.

The Eclipse fender was adopted early as a safety device. It would protect a pedestrian who had been struck by a streetcar. The leading element would pivot to a horizontal position if a person were struck and would fall onto it; the vertical spring steel cushion strips would prevent the person from striking the rigid car front. The Eclipse safety fender was used by many systems in the United States and was an integral part of Cleveland's streetcars.

The Eclipse fender was universal until the 4000- and 5000-series streetcars were built, starting in 1927. It was then superseded by the Eclipse wheel guard; a gate at the leading end of a car would be tripped if it struck a person or other object, triggering a "basket" that would drop onto the rails. It was used on 4000s, 5000s, 450–459, a few 150s, and the Presidents' Conference Committee (PCC) cars.

A safety fender of an approved type was required by municipal ordinance in Cleveland and most other places. A flanged steel wheel would easily cut off a person's limb should he or she be run over, so it was deemed imperative to have a safety fender. In the heyday of streetcars, pedestrian traffic was far heavier than today. This need for safety was met by the technology of the time: the Eclipse fender or wheel guard.

Electric track switches were activated by the trolley pole passing under a "pan" immediately in front of a facing point turnout. "Power on" threw the switch to the right, while "power off" threw it left. This device was used in most cities.

Early traction motors were large, requiring streetcars to have large wheels in the range of 30- to 36-inch diameter. Wheels of 32–34 inches were provided in all cars through the 1000–1052 series (except 1048) of 1913. After that, smaller motors of equal power (40–50 hp each) allowed lower-floor cars with consequent lower and/or fewer steps. Car 1048 and cars from the 1100 series of 1914 through the PCC cars of 1945–46 had 26-inch wheels.

The Breda LRVs have 28-inch wheels, a size commonly used on heavy-rail rapid transit cars and in keeping with the large, powerful monomotor used in that car. That motor is mounted longitudinally and drives both axles through a right-angle gear-unit on each end of the motor.

The classic streetcar had motors parallel to the axle, with each motor driving one axle through single-reduction spur gears.

The PCC streetcars had their 55-hp motors mounted at right angles to the axle, driving through a right-angle hypoid gear-unit.

Cleveland's several early street railway companies employed contemporary technology, and it appears from photographs that they employed a variety of cars from several sources. A fundamental change from short four-wheel cars to double-truck eight-wheel cars occurred in the 1890s, and cars grew larger in the first decade of the twentieth century until 50-foot cars became the norm about 1906–12. That appeared to be the longest practical size for city cars. Some interurban cars of 61 feet, 6 inches were operated into Cleveland over street railway lines by the Lake Shore Electric Railway and the Cleveland Southwestern Railway and Light Company. Other interurban operators used shorter cars, generally 50 to 55 feet in length.

After the consolidation that produced the Cleveland Railway Company, a classic period occurred in which streetcar development in Cleveland proceeded in a steady evolutionary pattern and culminated in a system that brought the art and science of hauling people down streets to a pinnacle that was seldom equaled and never surpassed during the period 1914 to 1945.

### Track

Cleveland track was standard gauge, 4 feet, 8½ inches. CR adopted high-quality track standards and was an industry leader in developing efficient methods to build and rebuild street railway track. The company was an innovator in developing or acquiring specialized equipment.

Eighty-pound ASCE rail laid on wooden ties with 12 inches of crushed stone ballast was the type of construction built on a private right-of-way where the grading consisted of cuts and fills.

Track work on Lorain Avenue is under way in 1938. CR compressor 0966 is to the right; a thermit welder is in the left foreground. *(Cleveland Railway photo, Blaine Hays Collection)*

Tracks formerly paved with brick were repaved with granite, and the sides of the street were paved with brick in certain cases.

In later years, in the 1920s, rail joints in paved track began to be welded in place with the Thermit process. Where girder rail was used in medians, welded joints were employed. Generally, Cleveland Railway followed the state-of-the-art of the street railway industry.

In several cases, girder rail was used on private rights-of-way, Clifton Boulevard's side-of-the-road arrangement being one. Others were the medians of Mayfield and Brookpark roads, which were designed to be paved when appropriate. Mayfield was paved in the late 1930s, whereas Brookpark between State Road and the West 33d Street loop was not paved until 1950.

Certain private rights-of-way, such as Cedar Hill and Euclid Heights Boulevard, used ASCE "T" rail. There was relatively little private right-of-way in Cleveland Railway's system, and only part of that had "T" rail.

Unlike railroads that constantly maintained track, CR occasionally rebuilt track, but it did not perform routine lining and tamping on private right-of-way ballasted track. The light axle loadings of streetcars and high-quality track construction made it unnecessary.

Most of Cleveland's street railway lines were double track, and the entire street was paved with granite block. In special local situations or as required by the municipality, brick, mass concrete, or asphalt sheet was used. Granite block was everlasting. When track was replaced, block was removed by first extracting a few blocks by jackhammering the joints, removing several, and then inserting the point of the pavement plow. The plow was a four-wheeled rail cart loaded with scrap iron that was towed by a work car. Blocks flew. Workmen piled them at the curb, and while rails were being replaced, laborers chipped off the old mortar. After new track was laid, blocks were replaced, new mortar broomed into the joints, and a new surface from the old returned to use. Many Cleveland streets still have granite block and street railway track under several layers of asphalt. It's a solid foundation.

A concrete pavement breaker was used where appropriate. It, too, was a four-wheeled rail cart with a heavy weight, like a pile driver, lifted by an electric motor geared to a mechanism that raised and dropped the weight.

Work cars carried away debris and brought material to work sites. The 0500 series dump cars built by the Differential Car Company of Findlay, Ohio, operated in three-car trains (motor-trailer-motor). Being double-ended, they could cross over on temporary crossovers ("bumpity switches" to school kids) and return to the shop. "V"-bottom dump cars of the 0600 series were sometimes used.

Differential crane cars of the 0700 series and flat trailers brought rails and set them. A welding car treated rail joints. A line car provided the means to work on trolley wire. Rail grinder 0800 removed corrugations, mill scale, or concrete from newly laid rails.

A number of other cars were built for special purposes. CR was well equipped.

### *Harvard Shops*

The foundation of Cleveland Railway's excellent maintenance was Harvard Shops, built in 1915 to replace the old Lakeview Shops facility, which was both too small and obsolete. The new shop was located at East 49th Street and Harvard Avenue in the southeast part of Cleveland. Its ten buildings were equipped with the latest technology, and the facility was capable of servicing 1,500 streetcars a year.

The Thew Shovel operated from contact with the overhead trolley wire and performed many jobs on track construction projects. *(Cleveland Railway photo, Blaine Hays Collection)*

At Harvard Shops, car body of 1026 is lifted by hoist as the truck is removed for motor repair. To the left is a single truck flusher and to the right is CER wooden snow sweeper 0105. *(Cleveland Railway photo, Blaine Hays Collection)*

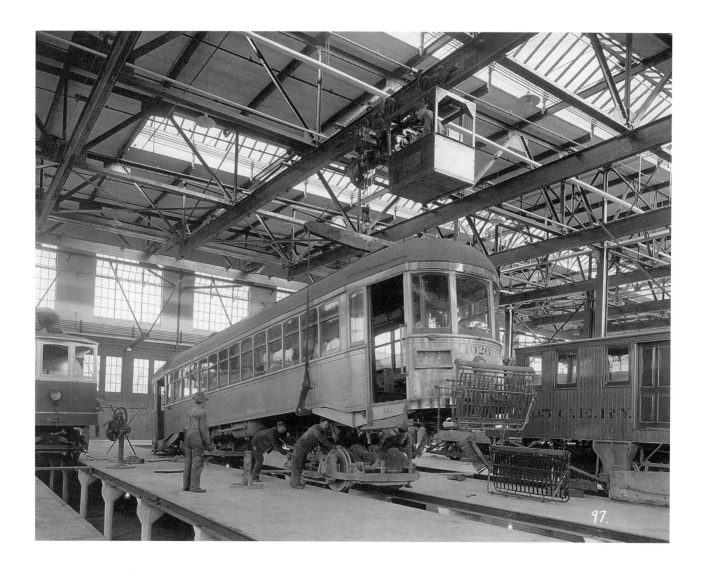

10 · CLEVELAND'S TRANSIT VEHICLES

Harvard Shops served CR and successor CTS until the end of streetcar operation in 1954. Several of its work cars helped to build the original Windermere–West 117th Street rapid transit line. By the mid-1950s, Harvard Shops no longer had a function to perform, but it still had one last contribution to make to rail transit in Cleveland.

In the mid-1950s, CTS sold Harvard Shops and the forty acres of valuable urban industrial land to the City of Cleveland for use by the Public Works Department. The sale raised capital to build the West 117th Street–West Park rapid transit extension in 1957–58.

That was before the Urban Mass Transportation Act of 1964, which authorized the first federally aided capital funding for transit. In the 1950s, there was only local funding, and CTS converted one asset it no longer needed, Harvard Shops, to a new asset that it wanted, the West Park extension. Thus its transit value lives on today, indirectly having been provided by the shareholders of the Cleveland Railway Company in 1912–15.

### Rapid Transit on a Shoe String: The Technology That Created the Shaker Heights Rapid Transit

When the Van Sweringen Company contracted with the Cleveland Railway Company to create what came to be known as the Shaker Heights Rapid Transit, full use was made of the physical plant and operations capability of CR.

In 1913 CR built the first track into the Van Sweringen domain. The new track ran from its Fairmount Boulevard line south along Coventry Boulevard to Shaker Boulevard and then east to Fontenay

SHRT car 58 is traveling east on the westbound track while single tracking operations are underway. OX is pulling a hopper car full of ballast as part of ongoing track maintenance. (Blaine Hays photo)

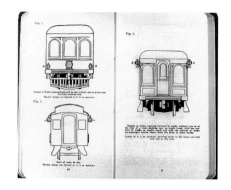

Road. Cleveland Railway also provided the operating equipment, the power, and the crews. Even in 1920, after the Van Sweringen Company had built the substructure (bridges, stations, track, and catenary) for its new express route to downtown Cleveland, the company leased the cars, purchased power as used, and contracted for operation and maintenance. CR even did all the accounting work. The Van Sweringens' strategy was a good way to conserve capital and get things running quickly.

The rapid transit's car fleet came from Cleveland Railway's 1200 series. The WH340 motor was standard for rapid transit and city cars. Standardization was not altered. The cars originally had Brill 51-E trucks of 4-feet-10-inch wheelbase, K-type platform controllers for single-car operation, and a coupler at the rear to pull a trailer. The Railway Company swapped trucks with newer cars having Brill 68-E-1 trucks of 6-foot wheelbase to provide a smoother ride at 45 mph on bolted ballasted track. The following cars gave up their 68-E-1 trucks for older 51-Es: 150–164 (fifteen cars) built by Cleveland Railway in 1924–25 and 484-499 (sixteen cars) built by Kuhlman Car Company in 1920.

Several cars received Standard Steel pressed steel trucks with 6-foot wheelbases, namely 3, 6, 8, 10, and 14 (originally 1203, 1206, etc.). Cars 1301–10 also had Standard Steel trucks when built by the railway.

In deference to Shaker riders' dislike for longitudinal seating, the cars' original benches along the left side were replaced with transverse seating. The Railway removed the transverse seating from 1140–49, 1299, and 1300, replacing it with benches to match 1152–1199. Other sources of transverse seating would have also been necessary, because the cars cited would have been insufficient to reequip 1200–1235 for Shaker use.

The cars were converted for multiple-unit HL control, and a field shunt was added, which raised running speed from 28 mph to 45 mph.

Cleveland Railway cars were converted as follows:

1200–1214 in 1920–22 had short jumper cables adjacent to the couplers for multiple-unit connection (later renumbered 9, 1–8, 10–14).

1209 rebuilt to cable (one end) and buttons (other end) and later to all buttons (later renumbered 36).

1215–1232 in 1925–27 had button-type electrical couplers (later renumbered 15–32).

1233–1235, trailers, retained 51-E trucks; had a button coupler at front end only (later renumbered 53–55).

56–58, trailers, former 2300s, obtained from CTS in 1945; had cable jumpers and retained longitudinal benches, coupler at front end only; 57 was never used in revenue service but was the yard office at Warrensville Center Road terminal.

4050, 4051, converted for one-man operation; operated for a time on the Shaker Rapid in the early 1930s.

From available evidence it appears that the first three cars (1200–1202) were not converted to multiple-unit control and these three were returned to street service in 1922, after having been renumbered. Car 1203 was converted to a multiple unit in 1922.

Cleveland Railway provided the Cleveland Interurban Railroad with electrical power. Shaker's new catenary had no substation of its own. Feeder cables connected Shaker's catenary to Cleveland Railway's 600-volts DC network.

The substation at the end of the railway's Woodland Avenue line near Shaker Square contributed power. Feeders on East 93d Street (East 105th Street crosstown line), East 55th Street, and other places all helped. It was not until the demise of streetcar and trolleybus service in Cleveland that the Shaker Rapid owned its own substations.

Shaker Rapid purchased the Woodland Avenue substation, with its two Westinghouse 1,500 kw units, along with a used General Electric 1,000-kw converter that had been installed in a room built into the piers of the Warrensville Center Road bridge across Shaker Boulevard. Shaker continued to buy power from CTS at East 55th Street and in the joint area from East 55th Street to the Cleveland Union Terminal. The downtown Shaker Rapid terminal was supplied from the CTS substation just east of the Cleveland Union Terminal.

In 1968, two 1,500-kw substations were constructed by Shaker Rapid to replace the obsolete and deteriorated Woodland Avenue facility. They were the latest solid-state technology and have served very well for over twenty-five years. One is at the end of the Van Aken line, and the other is near the Buckeye/Woodhill intersection, about midway between East 55th Street and Shaker Square. The GE unit under the Warrensville Center Bridge was kept operable as a standby and for extra-heavy crowds until it was retired and scrapped when the bridge was rebuilt. The Regional Transit Authority (RTA) added a 3,000-kw substation at Shaker Square and a similar 3,000-kw unit east of Warrensvile Center on Shaker Boulevard.

Light running maintenance was done at the CR's East 34th Street carhouse. Heavy maintenance and overhauls were done at Harvard Shops. In 1920 there was no need to invest in a shop for Shaker's exclusive use.

Some major changes occurred in 1930 when the rapid moved into the Cleveland Union Terminal. Its own shop was built at Kingsbury Run, near East 55th Street. The Cleveland Interurban Railroad Company (CIRR) continued to lease the 1200-type cars, but it no longer depended on the railway for operations.

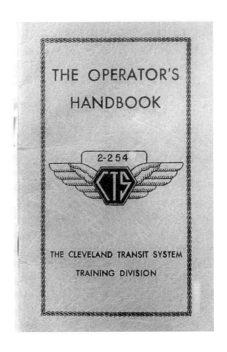

During the years when CR operated the Shaker Rapid, trolley wheels were used, the same as on city cars. A three-car train used only the front trolley because electric switch pans were located for single-motor cars. This put a big load on one trolley wheel. A 600-volt trainline carried power to trailing cars.

After CIRR moved into the Union Terminal, it became possible to use more than one trolley. There had been problems with burnt wheels and pitted wire, so the superintendent of equipment, Bill Powellson, equipped a five-car train with Miller Trolley Shoe Company steel shoes. It made a round trip from Warrensville Center Road and Moreland Boulevard using three trolleys, on cars 1, 3, and 5. It did not draw an arc. CIRR then converted all cars to shoes. Steel shoes were not self-lubricating like carbon, so it was necessary for line car 101, and later 1231, to apply graphite grease to the trolley wire at intervals. In 1967 Shaker converted to a shoe made of bearing bronze to reduce wire wear, putting the wear on the shoe instead of the nearly worn-out wire. This served to prolong the life of the overhead until the Greater Cleveland RTA modernization program of 1980.

CIRR had an electric switch at the Shaker Square junction. The pan was back a train length so that the lead car could throw the switch; and even if a trailing trolley should accidentally activate it, the motorman could stop and throw it manually. Shaker followed CR practice: power to the right, coast to the left.

The influence of Cleveland Railway on the Shaker Rapid was pervasive and generally to the benefit of all concerned. The Shaker system operated pretty much as built until the major modernization program of the 1980s. That it survived until the RTA era was a tribute to its original designers.

### The Trolleybus System

Trolleybus service began in Cleveland with the 4.6-mile Hough Avenue–East 105th Street line, which replaced the Payne Avenue streetcar line on May 1, 1936. This line was reported to be a milestone, since it was the first trunk trolleybus line to serve the downtown of a metropolitan area of a million residents. At that time trolleybus lines were either in small cities or merely provided crosstown or feeder service in large cities.

The twenty new Pullman forty-seat coaches had two 50-hp GE motors, one driving each wheel. There was no differential, a typical design of the time. The service proved popular, attracting sufficient new riders to cause the Cleveland Railway to buy eight more identical coaches the following year.

The railway announced that the Wade Park–East 118th Street line would also be converted to trackless trolleys, as they were usually called

Trackless trolleys 1201 and 1212 are being loaded aboard flat cars at the old Food Terminal siding. The retired TTs are on their way to a second life in Mexico City. *(Ross Barnard photo)*

in Cleveland. The Wade Park line shared with the Hough line the same inner route on Payne and Superior avenues. CR was unable to finance the conversion, however, so it temporarily used motor buses on Wade Park. This led to the use of bus trailers on Wade Park during World War II to cope with wartime traffic.

In 1942 the railway purchased from Twin Coach, the fifty-eight-seat, three-axle, vertically articulated Super-Twin trolleybus number 6000. It became known as the Queen Mary. Number 6000 had a single motor, a propulsion system that had become standard by this time. The coach had been demonstrated in 1940. In Cleveland it spent its entire career on Hough Avenue and was the only prewar articulated trolleybus in the United States. Chicago had a postwar version of different appearance. The use of an articulated trolleybus and motorbus trailers on Payne Avenue reflected CR's views as to how large volumes should be handled.

No other articulated trolley buses were obtained, in part because both the front and rear axles steered in opposite directions. The coach pivoted about the center axle and was rigid in the horizontal plane. As a result, the rear swung toward the curb as the coach pulled away from a stop. At times this caused problems.

Following the city's acquisition of Cleveland Railway in 1942, several plans were drawn up for conversion of the streetcar system to various combinations of rapid transit, trolleybus, and motorbus. The final choice was to build the Windermere–West 117th Street rapid transit line. Heavily patronized local services within the city were converted to trolleybus; the portions of these trunk lines beyond the city limits became express bus lines, and some later became feeders to the CTS rapid.

The 5.6-mile Wade Park line was officially converted to trolleybus on December 16, 1945. At first, only a few trolleybuses were in service; but as they became available, more were added until about twenty-four electric coaches were in service.

Deliveries continued as other lines were converted:

4.5-mile Cedar Avenue–East Boulevard, June 16, 1946
7.1-mile Broadway, August 16, 1947
8.8-mile East 105th Street–Miles Avenue crosstown, January 31, 1949
6.8-mile Union Avenue, January 31, 1948
5.4-mile Buckeye Road, July 1, 1948
5.7-mile Woodland Avenue, 1949
5.7-mile Clifton Boulevard–West 117th Street, 1949
6.2-mile Fulton–Denison–West Seventy-third Street, 1949
7-mile Kinsman Avenue, March 26, 1950
4.4-mile Detroit–West 117th Street, August 24, 1951
6.9-mile St. Clair–East 152d Street, November 3, 1951
6.5-mile Lorain Avenue–West 140th Street, June 14, 1952
6.5-mile Superior Avenue–East 129th Street, March 20, 1953

On April 27, 1952, the Euclid Avenue line, long home to the articulated streetcars, was converted "temporarily" to motorbus, but CTS announced that it would eventually be converted to trolleybus. The final order for new trolleybuses, Marmon-Harrington 1275–1324, was put into service on East 105th Street in 1951–52. Although the vehicles had destination signs for Euclid Avenue, the conversion was never implemented. At its peak in 1953–55, CTS's trolleybus system had 461 coaches operating on fifteen routes.

### *Technology*

All trolleybus lines had been streetcar lines. Use of the existing 600-volt DC power system allowed conversion to be done economically. Generally, the existing poles and feeder cables were used. New negative trolley wire was added. In some cases, new positive wire was erected in addition to the existing streetcar wire where the latter's location in the center of the street was not suitable for the curb-loading trolleybus, or when the existing wire was too worn. The streetcar wire was later removed. Trolleybus intersection specialwork was erected above existing streetcar wire. At conversion time, streetcar wire was cut down overnight.

Existing rotary converter substations continued in use. Existing rails provided negative return. Before the streets were repaved, CTS crews welded (or brazed) copper cable around broken rail joints. In some cases, curved rails were removed to allow restoration of the street crown. A copper cable was installed in its place to ensure negative continuity. The streetcar power supply continued in use even after the streets were repaved with several inches of asphalt.

In a few cases, such as Fulton, the street was narrow, so only a new negative wire was erected. The old round streetcar positive wire continued in use. Trolleybus wire was normally grooved, the hangers grasping the groove, allowing a smooth surface for the trolley shoes. Older streetcar wire was round, with hanger ears grasping around the wire. It worked well enough with trolley wheels but required that the ears be filed smooth to remove pits and burrs before shoes could be used. As an expedient, it worked.

In nearly all cases, existing support poles were used. However, at intersections having trolleybus overhead specialwork (turnouts and crossings), the much heavier weight required installation of heavy tubular poles bolted down to sturdy concrete foundations.

In one case, on Clifton Boulevard, from West 95th Street to West 117th Street, new poles with bracket arms were installed on the newly built concrete curb lanes that replaced the side-of-the-road streetcar private right-of-way. New positive and negative feeders were installed on this segment.

Another special case was West 117th Street, which had trolleybus wire erected for the Clifton and Detroit lines. A loop was provided for the Detroit line in the yard of Madison Car House at West 117th Street and Madison, which became the regular terminal for the Detroit line. Clifton deadhead coaches used new wire on West 117th to reach Madison Car House. Interestingly, West 117th was a proposed but never-built crosstown streetcar line. There was an existing pole line on one side of the street for feeders between Clifton and Detroit. There was a single streetcar track between Detroit and Madison for nonrevenue use, so there were poles on both sides of the street and feeders on this segment. Little new construction was required.

After the West 117th Street sewer explosion in 1953, CTS was forced to abandon Madison Car House. Temporary wire was hastily erected on Coutant Street (West 118th) between Clifton and Madison with a connection at Detroit so coaches could reach Denison Station. A sycamore tree was used for a pull-off. The segment between Clifton and Detroit was single track with a spring switch at each end. Wire was also erected on West 117th Street between Madison and Lorain, about a mile, to allow Clifton and Detroit coaches to reach Denison Station via Lorain and Denison avenues. There was already a pole line with feeders on one side of West 117th Street.

On the west side of town there was also deadhead wire on West 73d Street between Lorain and Denison, and there was also wire on Denison between Lorain and Ridge Road for Lorain coaches to reach Denison Station. There was no deadhead wire on the east side.

Trolleybus operating stations were: Hough, Superior, St. Clair, Fairmount, Woodhill, Madison, and Denison. It was common practice in other systems that converted streetcar lines to trolleybus to use as much of the existing plant as possible. Many were intended as a one-generation system to use up the remaining service life in an otherwise sunk investment. From CTS's action it is concluded that CTS also intended a one-generation trolleybus system. The final conversions of streetcar lines to trolleybus were facilitated by acquiring two fleets of used trolleybuses, some from Louisville, Kentucky, and some from Providence, Rhode Island, from transit systems that had already abandoned that mode. The trend had begun.

Cleveland's trolleybus lines lasted only until 1963, an eleven- or twelve-year service life for the last trolleybuses, the 1275–1324 Marmons.

CTS purchased fifty diesel buses each from General Motors (3500s), Mack (2700s), and White (2900s) concurrently with Marmon trolleybuses 1275–1324 in 1951–52. CTS kept careful accounting records of fuel consumption and maintenance costs. This led to subsequent purchase of GMs in fleet quantities to carry out conversion of the final streetcar lines in the early 1950s, for replacement of older motorbuses, and, in the 1960s, to replace the trolleybuses.

Other forces were then at work. Ridership declined greatly after 1948. The number of trolleybuses needed to meet schedules declined markedly. Maintenance of the large fixed plant was spread over fewer vehicle miles, thus raising unit costs.

One example is the Wade Park–East 118th Street line, converted with twenty-four coaches in 1945. Requirements had dropped to about eight coaches in the peak and three in the base at the time of conversion to motorbus.

In the early 1950s, the State of Ohio allowed municipal transit systems to use "city" license plates on their motor buses instead of commercial bus plates. This saved a significant sum. Commercial plates cost several hundred dollars (memory says four hundred dollars) versus a very small filing fee for city plates. Trolleybuses, not being "motor vehicles," did not need plates. This change in costs removed one advantage the trolleybus had.

The net result was that the fifty-passenger diesel hydraulic motor bus and, in particular, the General Motors bus, with its very reliable two-cycle engine, became the transit vehicle of choice for CTS.

The conversion process was completed on June 14, 1963. CTS removed and scrapped copper feeder cables, copper and phosphor-bronze trolleywire, and substation equipment that contained much copper. Significant funds were received that could have been applied to acquiring new motor buses. This all occurred before federal capital funding became available for transit systems. Some 1200–1274 series Marmon-Harrington trolleybuses were sold to Mexico City for further use.

Many poles remain along Cleveland streets to indicate where streetcars and successor trolleybuses once ran. Black tubular steel poles, many with malleable iron collars to counter corrosion at the ground line, remain in position, holding up streetlights, traffic signs, or nothing in particular. They are removed only when there is a reason to do so. Some octagonal concrete poles remain from the World War I period and immediate postwar period when a shortage of steel caused Cleveland Railway to use that type of pole. These are visual symbols of a "sunk investment" on the part of shareholders in the Cleveland Railway Company. Some substation buildings also remain, though they are now used for other purposes. There is evidence throughout Cleveland that shows where electric transit vehicles once plied the streets.

Cleveland's transit vehicles were not immune to accidents. The following photos show three of the most extensively damaged vehicles. Streetcar 316 *(below)* shows the results of a 1947 collision. Brand new bus 3184 *(overleaf, top)* was in a serious accident in 1949. Airporter car 170 *(overleaf, bottom)* was destroyed by an electrical fire in November 1972; the flames were so intense that the car's steel frame was bowed. *(Blaine Hays Collection)*

CHAPTER TWO

# Roster of Cleveland's Transit Vehicles

This is a roster of nineteenth- and twentieth-century transit equipment of companies that served the City of Cleveland and surrounding areas. Its four sections include streetcars (passenger and work equipment), rapid transit cars (passenger and work equipment), trackless trolleys, and buses. Information comes from official files of the Cleveland Railway Company, Cleveland Transit System, and Regional Transit Authority as well as from notes and memories of persons close to the situation.

The roster is a collection of facts and memories about the thousands of vehicles that have provided public transportation services in Cleveland, largely between 1903 (when the companies were consolidated) and the city's bicentennial celebration of 1996. During this great industrial and commercial era, hundreds of millions of Clevelanders have used and/or are using these vehicles on a day-to-day basis in the mundane task of going to and from their jobs. These transit vehicles provided a steady heartbeat and controlled the flow of lifeblood through the city for a century and a half.

Also included are the comments of two men whose experience includes impressive careers in the transit industry. J. William Vigrass and Robert Korach have generously shared their knowledge. Information concerning the electric streetcars prior to the century's turn was culled from the extensive research and notes of the late Kenneth S. P. Morse and Harry Christiansen.

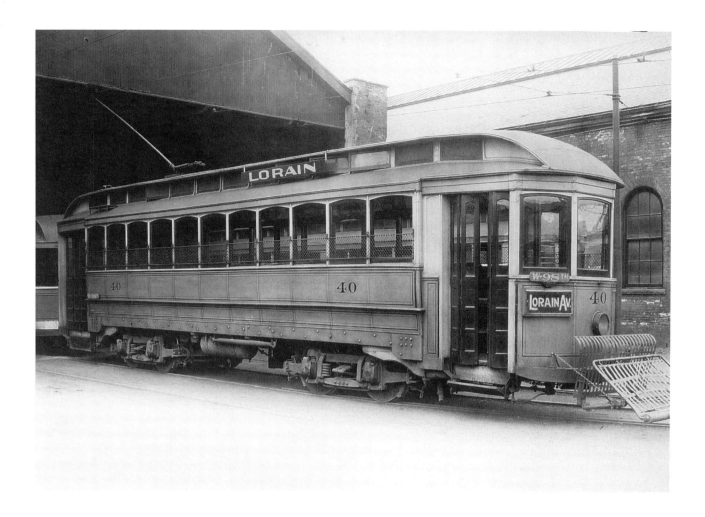

Car 40, a 28-foot box-type streetcar, at Lorain Station. *(Cleveland Railway photo, Blaine Hays Collection)*

From the start of the horse/cable car era to the consolidation of 1903, there was an entire generation of cars for which no information is available. It appears that passenger cars always had numbers—or, in rare case, names—on them. Freight cars and work equipment were originally identified by Roman numerals.

Most horse and cable car bodies became either passenger trailer cars or work cars (chiefly flat cars) during the electric era. Early work cars had very low numbers that were changed to higher numbers as the system grew (also adopted at that time was the practice of placing a zero prefix in work car numbering). Some of these early numbers survived on shop records and are included in the roster. Data on some cars are missing because those cars were either extensively rebuilt or scrapped before formal records were kept.

Some information has not been included in the roster due to space limitations, including data about automobiles and trucks and other nonrevenue vehicles used by the transit operators.

Names that appear in the roster, such as Apex, Bassicus, Cherry, Kricfalusy, Kubentz, Ours, Rearich, and Samworth, refer to salvage dealers of the era.

The stories of the vehicles as told in the roster lend personality to their tons of rolling steel and help illuminate the story of public transit.

The total number of known vehicles from the beginning of public transit in Cleveland to the time of the city's bicentennial celebration in 1996 is as follows:

| | |
|---|---:|
| streetcars | 2,730 |
| trackless trolleys | 461 |
| buses | 3,699 |
| rapid transit cars | 313 |
| Total | 7,203 |

Car 19, a semiconvertible-type streetcar, at Superior Station. *(Cleveland Railway photo, Blaine Hays Collection)*

PART A

## *Cleveland Streetcar Roster from the Time of Electrification*

The era of the electric streetcar in Cleveland began with construction of the Bentley-Knight system on Quincy, from East 46th Street, on July 26, 1884. This installation was not a success, but with the advent of successful electric operations four years later came a nationwide rush to improve and consolidate public transit routes.

Car 253, a nine bench–type open car, from preconsolidation days.
*(Cleveland Railway photo, Blaine Hays Collection)*

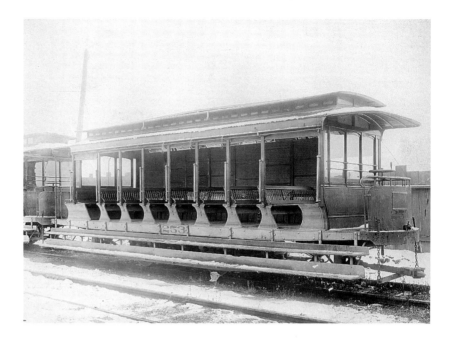

During the following two decades, 1,298 electric streetcars were constructed for Cleveland's lines by six manufacturers and two of the local transit companies. During this period, refining of the vehicles most suitable for Cleveland's needs took place. After much experimentation, two types emerged. First was the "box" car, rectangular in shape, of various sizes and seating plans, and with loading platforms on either end. Second was the "combination" car, with a door separating two car segments (open and closed sections, for example) that was controlled from the rear platform. Single-ended cars were clearly favored over double-ended cars throughout local traction history. Only a handful of cars, known as dinkeys, were double-ended.

All other car styles were rebuilt quickly to one of these configurations. These became standard, with longer car lengths the only change until 1914. Then newly appointed traction commissioner, Peter Witt, produced a design that changed street railway car design and operation for the duration.

### SECTION I
### *Outline of Electric Streetcars Owned by Predecessors of Cleveland Railway*

Here follows a simplified recapitulation of the 1,298 streetcars from the beginning of electrification to operation by the terms of the Tayler Ordinance (1910). Ownership of cars is stated for the consolidated systems after they acquired the underlier companies. Only the three surviving major systems that made up Cleveland Railway—the Cleve-

land Electric Railway Company (CER), the Cleveland City Railway Company (CCR), and the Municipal Traction Company (MT)—are used.

Existing data for these early fleets are incomplete. Order number and/or car number are provided when known. Fleets are listed in the order of their construction.

| Number of Cars | Year Built | Builder | System Built for | Order/ Car Series | Notes |
|---|---|---|---|---|---|
| 16 | 1889 | Brill | CER | order 2644 | 16 feet |
| 16 | 1889 | Brill | CER | | 16 feet |
| 10 | 1892 | Brill | CER | | 21 feet |
| 10 | 1892 | Brill | CER | order 3324 | 21 feet |
| 40 | 1892 | Brill | CCR | order 4366 | 21 feet |
| 6 | 1893 | Brill | CCR | order 4966 | SE, DR |
| 4 | 1893 | Brill | CCR | order 4966 | SE, DR |
| 5 | 1893 | Brill | CCR | order 4966 | SE, DR |
| 6 | 1893 | Brill | CCR | order 5239 | 20 feet |
| 21 | 1893 | Brill/Kuhlman | CCR | | J. A. Mehling CE, SE, combination |
| 15 | 1893 | Brill | CCR | | DR, SE, 9 bench open |
| 15 | 1893 | Brill | CCR | order 4978 | |
| 1 | 1893 | Brewer/Krehbiel | CCR | | articulated |
| 1 | 1893 | CER | CER | car 99 | Parlor Car became sightseeing car |
| 1 | 1894 | Brill | CER | order 5848 | DR, SE, 24 feet |
| 1 | 1894 | CCR | CCR | | "M. A. Hanna" Parlor Car, "Special" |
| 6 | 1895 | Brill | CCR | order 6610 | 21 feet |
| 8 | 1895 | Brill | CCR | | DR, SE, 9 bench, open |
| 8? | 1895? | Brill? | CCR | | RR, DT, 12 bench, open |
| 8 | 1895 | Brill | CCR | order 6256 | |
| 6 | 1896 | Brill | CCR | | DR, SE, 9 bench open |
| 6 | 1896 | Brill | CCR | order 7320 | 21 feet |
| 6 | 1896 | Brill | CCR | order 7131 | |
| 15 | 1896 | Brill | CCR | order 7087 | 10 bench, open |
| 58 | ? | Brill | CCR | | DT, open |
| 1 | 1898 | Brill | CER | order 8628 | |
| ? | 1898 | Brill | CER | order 8628 | DR, 28 feet closed |
| 1 | 1898 | Kuhlman | CER? | | Pay Car FN 500 |
| 1 | 1898 | CER | CER | | 4-casket funeral car combination car purchased from AB&C Interurban |
| 50 | 1899 | Stephenson | CER | | |
| 3 | 1899 | Stephenson | CER | | DR, 28 feet closed |
| 51 | 1899 | Brill | CER | order 9185 | DR, 28 feet closed |
| 50 | 1899 | Brill | CER | order 9185 | DR, 28 feet closed |
| 14 | 1899 | Brill | CCR | | RR, 30 feet closed |
| 38 | 1900 | Brill | CCR | | RR, 30 feet closed |

| Number of Cars | Year Built | Builder | System Built for | Order/Car Series | Notes |
|---|---|---|---|---|---|
| 50 | 1900 | Brill | CER | order 10167 | DR, 28 feet closed |
| 50 | 1900 | St. Louis | CER | | DT, DR, convertible |
| 18 | 1900 | Kuhlman | CCR | order 53 | DT, open |
| 10 | 1901 | Brill | CCR | order 11001 | DT, RR, convertible |
| 25 | 1901 | Jewett | CER | | DT, DR, convertible |
| 2 | 1901 | Brill | CER | | DR, 28 feet closed |
| 24 | 1901 | Brill | CER | order 10710 | DR, 28 feet closed |
| 1 | 1901 | CER | CER | car 817 | DR, 28 feet closed |
| 11 | 1901 | CER | CER | | DR, 29 feet closed |
| 48 | 1901 | Brill | CCR | | RR, 30 feet closed |
| 25 | 1902 | Kuhlman | CER | | RR, 30 feet closed |
| 28 | 1902 | Brill | CER | | RR, 30 feet closed |
| 88 | 1902 | CER | CER | | DR, 29 feet closed |
| 15 | 1902 | Stephenson | CER | | DT, DR, convertible |
| 15 | 1902 | Brill | CCR | order 11563 | DT, RR, convertible |
| 15 | 1902 | Kuhlman | CCR | order 126 | DT, RR, convertible |
| 10 | 1902 | Kuhlman | CER | | DR, 30 feet closed |
| 2 | 1903 | CER | CER | cars 815/816 | DR, 28 feet closed |
| 15 | 1903 | CER | CER | | DR, 29 feet closed |
| 25 | 1903 | Brill | CER | | RR, 30 feet closed |
| 25 | 1903 | Niles | CER | | 28 feet |
| 2 | 1904 | Brill | CER | | RR, 30 feet closed |
| 2 | 1904 | CER | CER | cars 1998/1999 | CE trailers |
| 53 | 1905 | Kuhlman | CER | order 263 | 45 feet convertible |
| 1 | 1905 | CER | CER | | 2-casket funeral car BLT new by CER |
| 63 | 1906 | Brill | CER | | |
| 22 | 1906 | Brill | CCR | | |
| 7 | 1906 | Brill | CER? | | |
| 47 | 1906 | Kuhlman | CER | order 300 | 45 feet convertible |
| 24 | 1906 | Stephenson | MT | cars 3326–3349 | wood, DR, DE |
| 50 | 1907 | St. Louis | MT | cars 3350–3399 | DR, semi-convertible |
| 26 | 1907 | Stephenson | MT | cars 3300–3325 | DR, 10-bench trailers from Brooklyn, N.Y. |
| 1 | 1908 | Cincinnati | MT | car 900 | 36 feet DR |

*Total:* 1,298 pre–Cleveland Railway cars

## SECTION 2
### CER and CCR after the Consolidation of 1903

On June 29, 1903, the Cleveland City Railway Company (CCR) was merged into the Cleveland Electric Railway Company (CER). The

Car 193, a rebuilt Mehling-type car, built at Lakeview Shops, rights sold to G. C. Kuhlman. *(Cleveland Railway photo, Blaine Hays Collection)*

Municipal Traction Company (MT) was not merged until formation of the Cleveland Railway Company (CR) on March 1, 1910.

During the first decade of the century, open cars were being rebuilt into closed cars, semi-convertible cars were being made permanently closed, and open-platform cars were being enclosed. CCR cars were renumbered into the fleets of CER within a few years after the consolidation. Longer closed cars were purchased during this period, and former CCR double-truck motor cars were rebuilt into trailers. This development was a reaction to tremendous increases in riding as the city grew in population and commerce. It was also an intentional action to eliminate the antiquated appearance of the fleet and to make the cars better fit a city fast developing.

Records of specific cars operating after the consolidation are scant. It is known that renumbered former CCR motor cars in the 100 and 200 series ran until scrapped in 1910, with a few lasting until 1914. CER trailers in the 300 and 600 series were scrapped between 1918 and 1929, so they operated through this period. CCR's 300-series motors and the dinkeys were discontinued between 1924 and 1939, and some of the Niles cars quit in 1916. By 1924 all Niles cars except some dinkeys were gone. The 400s and 500s lasted longer and operated well into the Cleveland Railway era, being scrapped between 1929 and 1931. Lower-numbered cars later worked into the fleet as 600s, 700s, and 800s lasted into the 1930s.

The Tom L. Johnson 3000s were used until the 1925–29 period, after which they were scrapped. Heavily rebuilt 125s and the 400s, which started as open cars, were used up through the 1915–19 period. While much of the renumbering of cars after 1903 has caused confusion as to what became what, information exists on when cars were retired, so it is possible to know the older equipment used by Cleveland Railway after 1910.

SECTION 3

## Cleveland Railway: Equipment Inherited from Predecessor Companies

The following is a list of the equipment received from the Cleveland Electric Railway Company and the Municipal Traction Company on March 1, 1910:

| Car Series | Total | Notes |
|---|---|---|
| 1–99 | 94 | missing 1, 3, 4, 5, 6 |
| 100–199 | 98 | missing 101, 112 |
| 200–299 | 94 | missing 280, 286, 294, 295, 298, 299 |
| 300–399 | 100 | |
| 400–499 | 100 | |
| 500–599 | 100 | |
| 600–663 | 64 | |
| 700–799 | 99 | missing 772 |
| 800–899 | 100 | |
| 900 | 1 | |
| 926–999 | 75 | |
| 3300–3325 | 26 | |

Of these 950 cars, 637 were closed cars and 313 were open. Of the total, 850 came from Cleveland Electric Railway and 100 from Municipal Traction Company.

NOTES ON MISSING CARS

1, 3, 4, 5, 6 • no such cars
101     burned on Clifton during strike on 5-19-08

Car 315, a trailer converted from a former CCR motor car. *(Cleveland Railway photo, Blaine Hays Collection)*

| | |
|---|---|
| 112 | CON to mail car 0204 on 4-3-08 |
| 280, 286, 294, 295, 298, 299 | • SCR prior to 1910 |
| 772 | collided with Pennsylvania Railroad train on Union 1-22-10 |
| 33, 210, 245, 259 | • SCR directly after takeover in 1910 |
| 262 | transferred as a passenger car but was soon REB to welder 024, SCR 1924. CR had two ST welders with number 024 at the same time, as the first one, former Parlor Car 99, was not SCR until 1925. |

DATES OF RETIREMENTS FOR CLEVELAND RAILWAY'S
INHERITED STREETCAR FLEETS

| | |
|---|---|
| 1911 | 283, 824, 3319 |
| 1912 | 200, 201, 203, 204, 206, 208, 212, 215, 217–222, 224, 226–229, 231–233, 235–244, 246–248, 251–253, 255, 257, 258, 275, 279, 285, 289, 291 (all ST open), 3300, 3301 |
| 1913 | 202, 374, 396 |
| 1914 | 2, 41, 64, 141, 205, 207, 209, 211, 213, 214, 216, 223, 225, 230, 234, 249, 250, 254, 256, 260–261, 263–267, 269–271, 274, 278, |

PART A: STREETCAR ROSTER · 29

|      | |
|------|--|
|      | 282, 284, 292, 293, 296, 297 (all ST open except first four) 3302–3318, 3320, 3325 |
| 1915 | 10–12, 14, 16, 17, 25, 26, 29, 32, 34, 38, 39, 47, 56, 73, 74, 78, 80, 83, 90–93, 96, 400, 401, 405, 412, 418, 430, 434, 439–443, 449, 452, 455, 456, 482–485, 488, 490–492, 494–496, 499, 814 (all DT former open cars except 814) |
| 1916 | 9, 15, 18, 21, 24, 28, 31, 35–37, 40, 44–46, 49, 57, 59–63, 65, 66, 69, 71, 85, 94, 95, 97–99, 369, 378, 404, 406–410, 414, 420, 422, 436, 438, 445, 453, 458, 460–464, 466, 468–479, 481, 486, 489, 493, 497, 498 (all DT former open cars except 369 and 378) |
| 1917 | 43, 151, 337, 424, 426, 447, 451, 465, 487 |
| 1918 | 19, 20, 22, 23, 27, 30, 42, 48, 52, 53, 55, 70, 72, 131, 159, 178, 330, 331, 338, 339, 349, 432, 457, 459, 505, 634, 670, 672, 674, 678, 686, 688, 696, 699 (670–699 series formerly 300–329 except 684) |
| 1919 | 147, 148, 150, 152–155, 160, 164, 167–169, 171, 172, 175, 334, 340, 344, 346, 348, 352, 676, 680, 682, 690, 694 |
| 1920 | 58, 59 (FN 446), 60 (FN 429), 149, 161, 166, 170, 173, 176, 180, 182, 186, 187, 190, 196, 480, 535, 594, 656, 692, 698, 969, 90 (FN 333), 92 (FN 336), 93 (FN 341), 94 (FN 343), 95 (FN 345), 98 (FN 354), 99 (FN 356), 9 (FN 366), 12 (FN 372), 14 (FN 380), 16 (FN 386), 17 (FN 390), 18 (FN 392), 19 (FN 394), 20 (FN 398) |
| 1921 | 898, Marcus A. Hanna Parlor Car, known as "Special" (no number) SCR Dec. 5, 1921 |
| 1922 | 23 (FN 355), 29 (FN 365), 50, 51, 52 (FN 428), 53 (FN 444), 55 (FN 448), 56 (FN 454), 57 (FN 467), 125–127, 132, 162, 179, 181, 183–185, 189, 194, 195, 199, 278, 673, 681, 683, 689, 691, 693, 736 |
| 1923 | 6 (FN 360), 7 (FN 362), 8 (FN 364), 10 (FN 368), 11 (FN 370), 13, 21 (FN 351), 22 (FN 353), 24 (FN 357), 25 (FN 358), 26 (FN 359), 27 (FN 361), 28 (FN 363), 30 (FN 367), 54, 61 (FN 416), 62 (FN 419) 63 (FN 421) 64 (FN 427), 65 (FN 402), 66 (FN 403), 67–68, 70 (FN 425), 71 (FN 433), 72 (FN 411), 73 (FN 413), 74 (FN 417), 75–77, 78 (FN 423), 79, 80 (FN 431), 81, 82, 83 (FN 435), 84, 85 (FN 437), 86–89, 91 (FN 335), 96 (FN 347), 97 (FN 350), 100, 102–103, 105, 128–130, 133–140, 142–146, 156–158, 163, 165, 174, 177, 591, 663, 671, 673, 675, 677, 679, 681, 683, 685, 687, 689, 691, 693, 695, 697, 736 |
| 1924 | 15 (FN 382), 31 (FN 371), 32 (FN 373), 33 (FN 375), 34 (FN 377), 35 (FN 379), 36 (FN 381), 37 (FN 383), 38 (FN 384), 39 (FN 385), 69 (FN 450), 106, 107, 113, 114, 120, 124, 603–604, 607, 609–611, 614 (FN 655), 615, 617–620, 625–626, 629–630, 635, 639, 641, 643, 647, 653, 657, 659, 661, 820, 822–823, 825, 827, 831, 833–834, 837, 844, 864–865, 868–870, 889 |

| | |
|---|---|
| 1925 | 711–726, 731–732, 735, 775, 777, 816–817, 895, 926–927, 929–930, 932–933, 937, 939, 944–948, 950–953, 966–967, 973, 997 |
| 1926 | 700–710, 717–730, 733–734, 738, 741, 743, 745, 748, 750, 752, 758–760, 762, 764, 766, 768, 770, 774, 776, 778, 780, 782, 786, 790, 792, 794, 796, 798, 815, 920 (FN 415), 931, 935–936, 938, 940–943, 949, 955, 957, 959–966, 970–971, 976–979, 995, 998 |
| 1927 | 632, 954, 963, 986–987, 996 |
| 1928 | 737, 739–740, 742, 744, 746, 754, 756, 784–785, 788, 956, 958, 964–965, 968, 972, 974–975, 980–985, 988–994, 999 |
| 1929 | 41 (FN 342), 502, 507, 509, 511–514, 516–519, 522, 525–526, 528–534, 546–549, 551, 580, 582, 585, 588, 593, 597, 601, 637, 640, 648, 651, 751, 755, 757, 761, 767, 771, 773, 777, 779, 781, 789, 793, 795, 818–819, 821, 826, 828–829, 836, 871, 872, 886–888, 891–894, 896, 897, 899 |
| 1930 | 747, 749, 753, 763, 765, 769, 779, 783, 787, 791, 797, 799 |
| 1931 | 40 (FN 332), 115–119, 121–123, 191, 192, 198, 500–501, 503–504, 505–506, 508, 510–511, 515, 520–521, 523–524, 527, 550, 552–579, 581, 583–584, 586–587, 589–590, 592, 595–596, 598–599, 600, 602, 605–606, 608, 612–613, 621–624, 627–628, 631, 633, 636, 638, 642, 644–646, 649–650, 652, 654, 655, (FN 614), 658, 660, 662, 664 (FN 314), 801, 812, 830, 832, 835, 838–843, 845–863, 866, 873–885, 890, 1998 (FN 7), 1999 (FN 8), 801, 812 |
| 1935 | 616, 804 |
| 1936 | 111, 188, 800, 802, 805–811 |
| 1940 | 43 (FN 388), 44 (FN 389), 46 (FN 393), 49 (FN 399), 867 |
| 1941 | 193, 42 (FN 387), 45 (FN 391), 47 (FN 395), 48 (FN 397), 803, 813 |

More data exist about the following series of cars, which were inherited by Cleveland Railway; hence, they are listed in numerical sequence.

### *Fleet Statistics: Deck Roof Cars, 500–599*

BLDR: G. C. Kuhlman; roof: deck; LEN: 500–551, 52'; 552–599, 50'; W: 50,120 lbs; seats: 47; CTL: 500–579, K6; 580–599, K35G; TRKS: 550–599, B27F; MOT: 550–577, 4WH101; 578–599, 4WH307F; BLT: 500–549, 1906; 550–599, 1905; CR T: 500–556, T 12; 557–569, T 13; 570–579, T 12B; 580–599 T 14. Individual differences noted as other subtypes below. SCR early enough that useful components were removed and stockpiled at Harvard for use on other equipment.

| | |
|---|---|
| 500 | CON OM 1920, TRKS B27F-E-1, MOT 4WH101B-6, SCR 9-17-31 |
| 501 | CON OM 1920, TRKS B27F-E-1, MOT 4WH101B-6, SCR 8-24-31 |

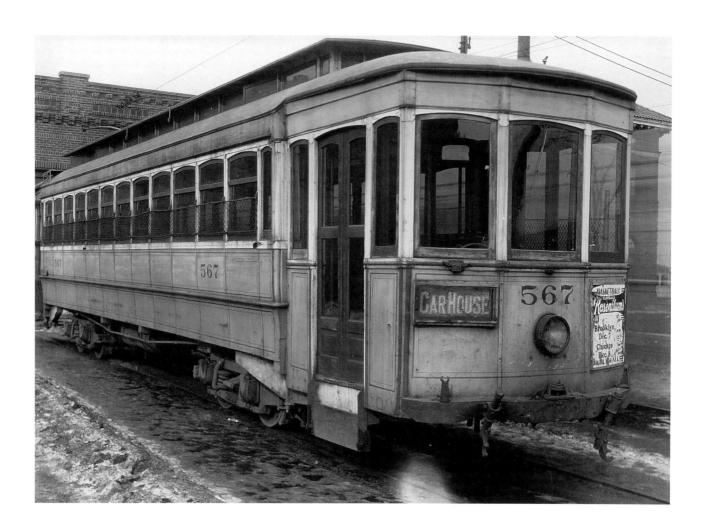

Car 567, deck-roof type, G. C. Kuhlman, 1905, CR type 13.
*(Cleveland Railway photo, Blaine Hays Collection)*

| | |
|---|---|
| 502 | SCR 8-16-29 |
| 503 | CON OM 1920, SCR 9-15-31 |
| 504 | REB 1910, TRKS B27F-E-1, SCR 8-24-31 |
| 505 | ACC, SCR 3-1-18 |
| 506 | CR T12C, REB 1914, CTL K35G, TRKS B27F-E-1, SCR 8-31-31 |
| 507 | TRKS B27F-E-1, SCR 8-19-29 |
| 508 | REB 1914, MOT 4WH101B-6, collision at 30th and Woodland 12-16-22, SCR 10-5-31 |
| 509 | REB 1913, MOT 4WH101B-6, TRKS B27F-1, SCR 8-13-29 |
| 510 | CON OM 1920, MOT 4WH101B-6, SCR 9-4-31 |
| 511 | REB 1914, MOT 4WH101B-6, TRKS B27F-E-1, SCR 9-11-31 |
| 512 | TRKS B27F-E-1, SCR 9-3-24 |
| 513 | TRKS B27F-E-1, SCR 9-28-29 |
| 514 | TRKS B27F-E-1, SCR 8-13-29 |
| 515 | REB 1914, CON OM 1920, TRKS B27F-E-1, SCR 9-21-31 |

| | |
|---|---|
| 516 | MOT 4WH101B-6, HDL Dayton 1583A, TRKS B27F-1, SCR 9-6-29 |
| 517 | CON OM 1920, TRKS B27F-E-1, SCR 9-6-29 |
| 518 | TRKS B27F-E-1, SCR 8-29-29 |
| 519 | SCR 9-3-29 |
| 520 | REB 1914, CON OM 1920, MOT 4WH101B-6, TRKS B27F-E-1, SCR 9-29-31 |
| 521 | REB 1914, CON OM 1920, MOT 4WH101B-6, TRKS B27F-E-1, collision 30th and Woodland 2-16-22, SCR 9-23-31 |
| 522 | TRKS B27F-E-1, seats 46, wrecked 12-21-20, SCR 8-16-29 |
| 523 | CR T 12 A, REB 1914, CON OM 1920, MOT 4WH101B-6, CTL K35G, TRKS B-27F-E-1, SCR 9-25-31 |
| 524 | MOT 4WH101B-6, TRKS B27F-E-1, SCR 10-1-31 |
| 525 | SCR 9-22-29 |
| 526 | CR T 12A, MOT 4WH101B-6, TRKS B27F-E-1, SCR 8-26-29 |
| 527 | REB 1914, CON OM 1920, TRKS B27F-E-1, SCR 10-7-31 |
| 528 | CON OM 1920, MOT 4WH101B-6, TRKS B27F-E-1, SCR 7-24-29 |
| 529 | TRKS B27F-E-1, SCR 7-22-29 |
| 530 | MOT 4WH101B-6, TRKS B27F-E-1, SCR 7-24-29 |
| 531 | MOT 4WH101B-6, TRKS B27F-E-1, SCR 1-12-29 |
| 532 | SCR 7-26-29 |
| 533 | MOT 4WH101B-6, TRKS B27F-E-1, SCR 7-26-29 |
| 534 | MOT 4WH101B-6, TRKS B27F-E-1, SCR 7-19-29 |
| 535 | wrecked at Woodhill Station, SCR 12-17-19 |
| 536 | MOT 4WH101B-6, TRKS B27F-E-1, SCR 9-24-29 |
| 537 | CR T12A, CTL K28B, SCR 8-29-29 |
| 538 | MOT 4WH101B-6, TRKS B27F-E-1, SCR 7-17-29 |
| 539 | MOT 4WH101B-6, SCR 8-6-29 |
| 540 | MOT 4WH101B-6, TRKS B27F-1, SCR 9-26-29 |
| 541 | TRKS B27F-1, SCR 8-6-29 |
| 542 | SCR 8-8-29 |
| 543 | TRKS B27F-1, SCR 8-1-29 |
| 544 | MOT 4WH101B, TRKS B27F-E-1, SCR 10-1-29 |
| 545 | SCR 8-8-29 |
| 546 | MOT 4WH101B, TRKS B27F-1, SCR 8-22-29 |
| 547 | SCR 7-30-29 |
| 548 | CR T 12A, MOT 4WH101B, CTL K28B, SCR 7-30-29 |
| 549 | MOT 4WH101B, TRKS B27F-1, SCR 7-19-29 |
| 550 | SCR 11-3-31 |
| 551 | SCR 9-18-29 |
| 552 | CR T 12B, SCR 11-3-31 |
| 553 | CR T 12B, REB 1914, SCR 10-30-31 |
| 554 | CR T 12B, SCR 11-5-31 |

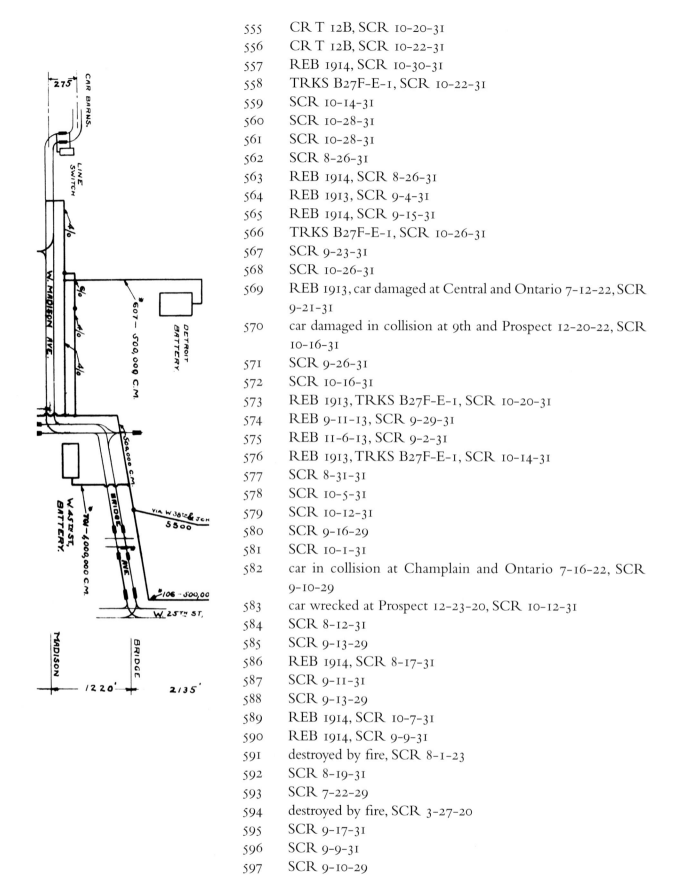

| | |
|---|---|
| 555 | CR T 12B, SCR 10-20-31 |
| 556 | CR T 12B, SCR 10-22-31 |
| 557 | REB 1914, SCR 10-30-31 |
| 558 | TRKS B27F-E-1, SCR 10-22-31 |
| 559 | SCR 10-14-31 |
| 560 | SCR 10-28-31 |
| 561 | SCR 10-28-31 |
| 562 | SCR 8-26-31 |
| 563 | REB 1914, SCR 8-26-31 |
| 564 | REB 1913, SCR 9-4-31 |
| 565 | REB 1914, SCR 9-15-31 |
| 566 | TRKS B27F-E-1, SCR 10-26-31 |
| 567 | SCR 9-23-31 |
| 568 | SCR 10-26-31 |
| 569 | REB 1913, car damaged at Central and Ontario 7-12-22, SCR 9-21-31 |
| 570 | car damaged in collision at 9th and Prospect 12-20-22, SCR 10-16-31 |
| 571 | SCR 9-26-31 |
| 572 | SCR 10-16-31 |
| 573 | REB 1913, TRKS B27F-E-1, SCR 10-20-31 |
| 574 | REB 9-11-13, SCR 9-29-31 |
| 575 | REB 11-6-13, SCR 9-2-31 |
| 576 | REB 1913, TRKS B27F-E-1, SCR 10-14-31 |
| 577 | SCR 8-31-31 |
| 578 | SCR 10-5-31 |
| 579 | SCR 10-12-31 |
| 580 | SCR 9-16-29 |
| 581 | SCR 10-1-31 |
| 582 | car in collision at Champlain and Ontario 7-16-22, SCR 9-10-29 |
| 583 | car wrecked at Prospect 12-23-20, SCR 10-12-31 |
| 584 | SCR 8-12-31 |
| 585 | SCR 9-13-29 |
| 586 | REB 1914, SCR 8-17-31 |
| 587 | SCR 9-11-31 |
| 588 | SCR 9-13-29 |
| 589 | REB 1914, SCR 10-7-31 |
| 590 | REB 1914, SCR 9-9-31 |
| 591 | destroyed by fire, SCR 8-1-23 |
| 592 | SCR 8-19-31 |
| 593 | SCR 7-22-29 |
| 594 | destroyed by fire, SCR 3-27-20 |
| 595 | SCR 9-17-31 |
| 596 | SCR 9-9-31 |
| 597 | SCR 9-10-29 |

Car 534, deck-roof type, G. C. Kuhlman, 1906, CR type 12. *(Cleveland Railway photo, Blaine Hays Collection)*

598    SCR 9-2-31
599    car in collision at E. 9th and Prospect 12-20-22, SCR 8-19-31

### Fleet Statistics: Railroad Roof, 600–664

BLDR: J. G. Brill; roof: railroad; LEN: 42' 9"; type: 30' box; seats: 38; crew size: 2-man; HDL: Dayton 348A; FEN: Eclipse; W: 44,600 lbs; TRKS: Brill 27F; CR T 5. These data apply to all except where noted. Cars 600–618 and 620–664 were "800-type." SCR early enough that useful components were removed and stockpiled at Harvard for use on other equipment.

600    BLT 1902, FN 604, CR T6C, REB 1910, MOT 4WH101, CTL K35G, SCR 9-4-31
601    BLT 1901, FN 490, REB 1911, TRKS B27F-E-1, CTL K6, MOT 4WH307F, SCR 9-10-29
602    BLT 1901, FN 548, REB 1910, CTL K6, MOT 4WH307F, SCR 9-29-31
603    BLT 1901, SCR 1924
604    BLT 1899, SCR 1924
605    BLT 1902, FN 596, REB 1910, MOT 4WH101, CTL K6, SCR 8-3-31

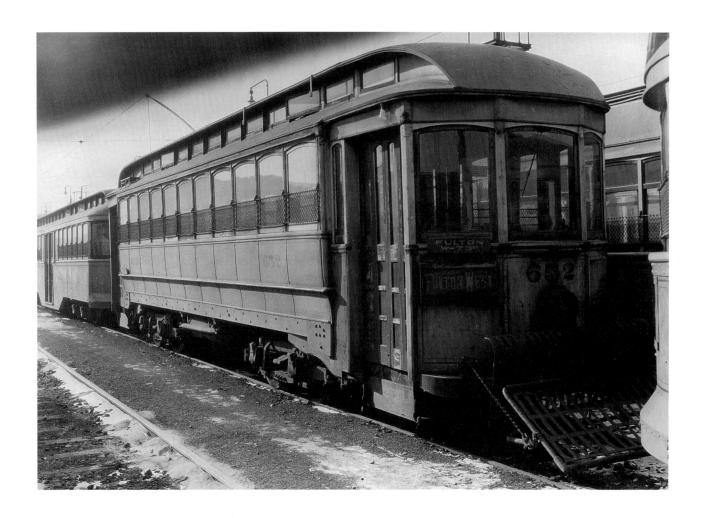

Car 652, deck-roof type, G. C. Kuhlman, 1902, CR type 5. *(Cleveland Railway photo, Blaine Hays Collection)*

| | |
|---|---|
| 606 | BLT 1901, FN 512, REB 1910, CR T6C, MOT 4WH101, CTL K35G, SCR 8-19-31 |
| 607 | BLT by Kuhlman in 1902, SCR 1924 |
| 608 | BLT 1902, REB 1910, MOT 4WH307F, CTL K6, SCR 9-2-31 |
| 609 | BLT 1899, SCR 1931 |
| 610 | BLT 1900, SCR 1924 |
| 611 | BLT 1900, SCR 1924 |
| 612 | BLT 1901, FN 490, REB 1910, MOT 4WH101, CTL K6, SCR 8-17-31 |
| 613 | BLT 1901, FN 518, REB 1910, MOT 4WH307F, CTL K6, SCR 8-10-31 |
| 614 | BLT 1899, REN 655 in 1924, SCR 1931 |
| 615 | BLT 1902, SCR 1924 |
| 616 | BLT 1901, FN 532, REB 1910, CR T 6C, MOT 4WH101, CTL K35G, stored since 1930, SCR 8-12-31 |
| 617 | BLT 1900, CR T 6, SCR 1924 |
| 618 | BLT 1900, CR T 5A, SCR 1924 |
| 619 | BLT by Kuhlman in 1900 from two horsecar bodies, CR T 4, |

| | MOT 4WH101B-6, CTL K10, TRKS B27F, became farebox storage shed at St. Clair Station, SCR 6-24-35 |
|---|---|
| 620 | BLT 1902, SCR 1924 |
| 621 | BLT 1902, FN 618, REB 1910, CR T 6, MOT 4WH101, CTL K6, SCR 8-24-31 |
| 622 | BLT 1901, FN 510, REB 1910, MOT 4WH307F, CTL K6, SCR 8-12-31 |
| 623 | BLT 1902, FN 616, REB 1910, CR T 6, MOT 4WH101, CTL K6, SCR 10-12-31 |
| 624 | BLT 1901, FN 584, REB 1910, CR T 6C, MOT 4WH101B-6, CTL K35G, SCR 10-14-31 |
| 625 | BLT 1900, SCR 1924 |
| 626 | BLT 1901, SCR 1924 |
| 627 | BLT 1901, FN 586, REB 1910, MOT 4WH307F, CTL K6, SCR 10-16-31 |
| 628 | BLT 1901, REB 1910, CR T 6C, MOT 4WH101, CTL K35G, SCR 10-20-31 |
| 629 | BLT 1900, SCR 1924 |
| 630 | BLT 1900, CR T 6C, SCR 1924 |
| 631 | BLT 1902, REB 1910, MOT 4WH307F, CTL K6, SCR 8-31-31 |
| 632 | BLT 1901, FN 496, REB 1910, CR T 6D, MOT 4WH101, CTL K28B, SCR 11-3-27 |
| 633 | BLT 1899, FN 396, REB 1910, MOT 4WH307F, CTL K6, SCR 7-1-31 |
| 634 | BLT 1900, wrecked in 1918, SCR 2-23-18 |
| 635 | BLT 1900, CR T 6, SCR 1924 |
| 636 | BLT 1902, FN 666, REB 1910, MOT 4WH307F, CTL K6, SCR 7-23-31 |
| 637 | BLT by Kuhlman in 1902, FN 664, REB 1910, MOT 4WH307F, CTL K6 |
| 638 | BLT 1901, FN 592, CR T 6C, REB 1910, MOT 4WH101B-6, CTL K35G, SCR 10-5-31 |
| 639 | BLT 1900, CR T 6C, SCR 1924 |
| 640 | BLT by Kuhlman in 1902, MOT 4WH307F, CTL K6, SCR 7-22-29 |
| 641 | BLT 1899, CR T 6C, involved in collision on Public Square 12-19-22, SCR 1924 |
| 642 | BLT 1902, FN 668, REB 1910, MOT 4WH307F, CTL K6, SCR 10-26-31 |
| 643 | BLT 1899, SCR 1924 |
| 644 | BLT 1901, FN 594, REB 1910, MOT 4WH307F, CTL K6, SCR 11-5-31 |
| 645 | BLT by Kuhlman in 1902, FN 644, REB 1910, MOT 4WH307F, CTL K6, SCR 11-25-31 |

| | |
|---|---|
| 646 | BLT by Kuhlman in 1902, REB 1910, MOT 4WH307F, CTL K6, SCR 11-3-31 |
| 647 | BLT 1899, CR T 6C, SCR 1924 |
| 648 | BLT by Kuhlman in 1902, CR T 6, became welder 0696 in 1929, SCR 9-11-47 |
| 649 | BLT 1901, FN 514, REB 1910, CR T 6, MOT 4WH101, CTL K6, SCR 9-9-31 |
| 650 | BLT 1901, FN 672, REB 1910, MOT 4WH307F, CTL K6, SCR 10-28-31 |
| 651 | BLT 1901, FN 576, REB 1910, MOT 4WH307F, CTL K6, SCR 8-29-29 |
| 652 | BLT by Kuhlman in 1902, REB 1910, MOT 4WH307F, CTL K6, SCR 10-7-31 |
| 653 | BLT 1900, SCR 1924 |
| 654 | BLT by Kuhlman in 1902, REB 1910, MOT 4WH307F, CTL K6, SCR 10-30-31 |
| 655 | BLT 1899, FN 398, REB 1910, CR T6C, REN 614 in 1924, MOT 4WH101, CTL K6, SCR 9-15-31 |
| 656 | BLT by Kuhlman in 1902, destroyed by fire at Harvard Shops 12-23-20, SCR 1920 |
| 657 | BLT 1900, CR T 6C, car sold in ? |
| 658 | BLT by Kuhlman in 1902, REB 1910, MOT 4WH101, CTL K6, SCR 8-26-31 |
| 659 | BLT 1900, car sold in ? |
| 660 | BLT by Kuhlman in 1902, REB 1910, CR T 6, MOT 4WH101, CTL K6, SCR 7-16-31 |
| 661 | BLT 1900, CR T 6C, car sold in ? |
| 662 | BLT 1902, FN 670, REB 1910, CR T 6C, MOT 4WH101, CTL K35G, SCR 11-23-31 |
| 663 | BLT 1900, involved in serious ACC, SCR 1923 |
| 664 | BLT 1903, FN 314, REN in 1915, CR T 5A, MOT 4WH307F, CTL K35G, SCR 7-8-31 |

### Fleet Statistics: Deck Roof, 700–799

BLDR: J. G. Brill; roof: deck; LEN: 41' 3"; type: 28' box; W: 44,600 lbs; seats: 38; crew size: 2-man; MOT: 2 WH101-B-2; HDL: Dayton 348A; FEN: Eclipse; CTL: K12; TRKS: Brill 27F. These data apply to all except where noted.

| | |
|---|---|
| 700 | BLT 1899, REB 1910, CR T 3, SCR 6-30-26 |
| 701 | BLT 1900, REB 1910, CR T 3B, CTL K10, MOT 2WH101B-6, SCR 6-30-26 |
| 702 | BLT 1899, REB 1910, CR T 3, MOT 2WH101B-6, SCR 6-28-26 |

Car 795, deck-roof type, J. G. Brill, 1900, CR type 1. *(Cleveland Railway photo, Blaine Hays Collection)*

| | |
|---|---|
| 703 | BLT 1900, REB 1910, CR T 3, SCR 6-30-26 |
| 704 | BLT 1899, REB 1910, CR T 3, MOT 2WH101, SCR 6-29-26 |
| 705 | BLT 1900, REB 1910, CR T 3, MOT 2WH101, CTL K10, SCR 6-26-26 |
| 706 | BLT 1899, REB 1910, CR T 3, MOT 2WH101B-6, SCR 7-2-26 |
| 707 | BLT 1900, REB 1910, CR T 3B, MOT 2WH101B-6, CTL K10, SCR 6-30-26 |
| 708 | BLT 1899, REB 1910, CR T 3, MOT 2WH101, SCR 6-28-26 |
| 709 | BLT 1900, REB 1910, CR T 3, MOT 2WH101B-6, SCR 1926 |
| 710 | BLT 1899, REB 1910, CR T 3B, MOT 2WH101B-6, CTL K10, SCR 6-28-26 |
| 711 | BLT 1900, CR T 3B, 2 MOT, SCR 1925 |
| 712 | BLT 1899, CR T 3B, 2 MOT, body sold, 1925 |
| 713 | BLT 1900, CR T 3B, 2 MOT, SCR 1925 |

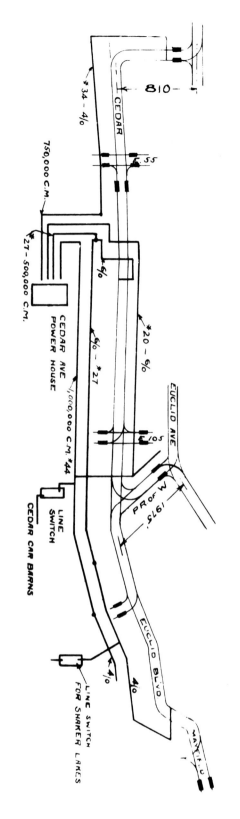

| | |
|---|---|
| 714 | BLT 1899, CR T 3, 2 MOT, SCR 1925 |
| 715 | BLT 1900, CR T 3B, 2 MOT, SCR 1925 |
| 716 | BLT 1899, CR T 3, 2 MOT, SCR 1925 |
| 717 | BLT 1900, CR T 3B, 2 MOT, SCR 1925 |
| 718 | BLT 1899, CR T 3, 2 MOT, SCR 1925 |
| 719 | BLT 1900, CR T 3, 2 MOT, SCR 1925 |
| 720 | BLT 1899, CR T 3, 2 MOT, SCR 1925 |
| 721 | BLT 1900, CR T 3, 2 MOT, SCR 1925 |
| 722 | BLT 1899, CR T 3, 2 MOT, SCR 1925 |
| 723 | BLT 1900, CR T 3B, 2 MOT, SCR 1925 |
| 724 | BLT 1899, CR T 3B, 2 MOT, SCR 1925 |
| 725 | BLT 1900, CR T 3B, 2 MOT, SCR 1925 |
| 726 | BLT 1899, CR T 3, 2 MOT, SCR 1925 |
| 727 | BLT 1900, REB 1910, CR T 3B, MOT 2WH101, SCR 8-5-26 |
| 728 | BLT 1899, CR T 3B, 2 MOT, SCR 1926 |
| 729 | BLT 1900, REB 1910, CR T 3B, MOT 2WH101, CTL K10, SCR 7-6-26 |
| 730 | BLT 1899, REB 1910, CR T 3B, MOT 2WH101, CTL K10, SCR 7-2-26 |
| 731 | BLT 1900, CR T 3B, 2 MOT, SCR 1925 |
| 732 | BLT 1899, CR T 3B, 2 MOT, SCR 1925 |
| 733 | BLT 1900, CR T 3B, 2 MOT, used by electrical department for a few years and REN 024, Probably SCR before 1930 |
| 734 | BLT 1899, CR T 3B, 2 MOT, CON to wrecker 0149 in 1910 and assigned to Cedar Car House, SCR 4-24-45 |
| 735 | BLT 1900, CR T 3B, 2 MOT, SCR 1925 |
| 736 | BLT 1899, CR T ?, CON to a trailer in 1914 and sold to South Bend, IN in 1922, REN 171 there |
| 737 | BLT 1900, REB 1910, CR T 1, MOT 4WH307F, CTL K35G, SCR 7-31-28 |
| 738 | BLT 1899, REB 1910, CR T 2A, MOT 4WH101, CTL K6, SCR 8-10-26 |
| 739 | BLT 1900, REB 1910, CR T 1, MOT 4WH307F, CTL K35G, SCR 8-7-28 |
| 740 | BLT 1899, REB 1910, CR T 2A, MOT 4WH101, CTL K6, SCR 7-24-28 |
| 741 | BLT 1900, CR T 1, REB to welder 0699 in 1928, SCR 4-28-41 |
| 742 | BLT 1899, REB 1910, CR T 2A, MOT 4WH101, CTL K6, SCR 10-31-28 |
| 743 | BLT 1900, REB 1910, CR T 1, MOT 4WH307F, CTL K35G, SCR 8-11-28 |
| 744 | BLT 1899, REB 1910, CR T 2A, MOT 4WH101, CTL K6, SCR 7-17-28 |
| 745 | BLT 1900, REB 1910, CR T 1, MOT 4WH307F, CTL K35G, SCR 8-17-26 |

746 BLT 1899, REB 1910, CR T 2A, MOT 4WH101B-6, CTL K6, SCR 10-2-28

747 BLT 1900, REB 1910, CR T 1, MOT 4WH307F, CTL K35G, SCR 4-17-30

748 BLT 1899, REB 1910, CR T 2A, MOT 4WH101B, CTL K6, SCR 8-19-26

749 BLT 1900, REB 1910, CR T 1, MOT 4WH307F, CTL K35G, SCR 4-10-30

750 BLT 1899, REB 1910, CR T 2A, MOT 4WH101, CTL K6, SCR 7-6-26

751 BLT 1900, REB 1910, CR T 1, MOT 4WH307F, CTL K6, SCR 8-8-29

752 BLT 1899, REB 1910, CR T 2A, MOT 4WH101, CTL K6, SCR 8-19-26

753 BLT 1900, REB 1910, CR T 1, MOT 4WH307F, CTL K35G, SCR 4-17-30

754 BLT 1899, REB 1910, CR T 2A, MOT 4WH101, SCR 6-5-28

755 BLT 1900, REB 1910, CR T 1, MOT 4WH307F, CTL K6, SCR 8-6-29

756 BLT 1899, REB 1910, CR T 2A, MOT 4WH101, CTL K6, SCR 10-9-28

757 BLT 1900, REB 1910, CR T 1, MOT 4WH307F, CTL K35G, SCR 8-19-29

758 BLT 1899, REB 1910, CR T 2A, MOT 4WH101, CTL K6, SCR 6-26-26

759 BLT 1900, REB 1910, CR T 1, MOT 4WH307F, CTL K35G, SCR 7-15-26

760 BLT 1899, REB 1910, CR T 2A, MOT 4WH101, CTL K6, SCR 8-5-26

761 BLT 1900, REB 1910, CR T 1, MOT 4WH307F, CTL K35G, SCR 7-17-29

762 BLT 1899, REB 1910, CR T 2A, MOT 4WH101, CTL K6, SCR 8-12-26

763 BLT 1900, REB 1910, CR T 1, MOT 4WH307F, CTL K35G, SCR 4-10-30

764 BLT 1899, REB 1910, CR T 2A, MOT 4WH101, CTL K6, SCR 8-17-26

765 BLT 1900, REB 1910, CR T 1, MOT 4WH307F, CTL K6, SCR 4-3-30

766 BLT 1899, REB 1910, CR T 2A, MOT 4WH101, CTL K6, SCR 8-12-26

767 BLT 1900, REB 1910, CR T 1A, MOT 4WH 307F, CTL K6, SCR 7-24-29

768 BLT 1899, CON to meter car 0152 in 1910 and SCR late in 1946

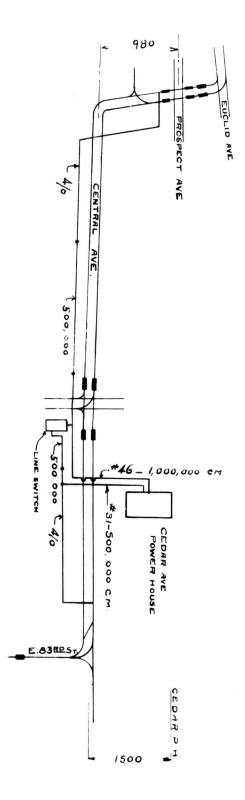

769  BLT 1900, REB 1910, CR T 1, MOT 4WH307F, CTL K35G, SCR 4-3-30

770  BLT 1899, REB 1910, CR T 2A, MOT 4WH101, CTL K6, damaged in collision on Broadway 8-6-21, SCR 8-17-26

771  BLT 1900, REB 1910, CR T?, MOT 4WH101, CTL K35G, SCR 8-16-29

772  BLT 1899, CR T 3, 2 MOT, destroyed by train on Union line 1-22-10, SCR

773  BLT 1900, REB 1910, CR T 1A, MOT 4WH307F, CTL K6, SCR 8-13-29

774  BLT 1899, REB 1910, CR T 2A, MOT 4WH101, CTL K6, damaged in collision on Broadway 8-6-21, SCR 8-10-26

775  BLT 1900, CR T 3C, wired for train operation in 1913, SCR 1925

776  BLT 1899, REB 1910, CR T 2A, MOT 4WH101, CTL K6, SCR 8-12-26

777  BLT 1900, CR T 3D, wired for train operation in 1913, SCR 1925

778  BLT 1899, REB 1910, CR T 2A, MOT 4WH101, CTL K6, SCR 8-5-26

779  BLT 1900, REB 1910, CR T 1, MOT 4WH307F, CTL K35G, SCR 4-10-30

780  BLT 1899, REB 1910, CR T 2A, MOT 4WH101B-2, CTL K6, SCR 7-17-26

781  BLT 1900, REB 1910, CR T 1A, MOT 4WH307F, CTL K6, SCR 10-3-29

782  BLT 1899, REB 1910, CR T 2A, MOT 4WH101B-6, CTL K6, SCR 8-10-26

783  BLT 1900, REB 1910, CR T 1, MOT 4WH307F, CTL K35G, SCR 4-17-30

784  BLT 1899, REB 1910, CR T 2A, MOT 4WH101, CTL K6, SCR 7-20-28

785  BLT 1900, REB 1910, CR T 1, MOT 4WH307F, CTL K35G, SCR 10-31-28

786  BLT 1899, REB 1910, CR T 2A, MOT 4WH101, CTL K6, SCR 7-6-26

787  BLT 1900, REB 1910, CR T 1, MOT 4WH307F, CTL K35G, SCR 4-3-30

788  BLT 1899, REB 1910, CR T 2A, MOT 4WH101, CTL K6, SCR 7-17-28

789  BLT 1900, REB 1910, CR T 1A, MOT 4WH307, CTL K6, SCR 7-30-29

790  BLT 1899, REB 1910, CR T 2A, MOT 4WH101, CTL K6, SCR 7-2-26

791  BLT 1900, REB 1910, CR T 1A, MOT 4WH307F, CTL K6, SCR 3-28-30

792  BLT 1899, REB 1910, CR T 2A, MOT 4WH101, CTL K6, SCR 8-17-26

793  BLT 1900, REB 1910, CR T 1, MOT 4WH307F, CTL K35G, SCR 8-1-29

794  BLT 1899, REB 1910, CR T 2A, MOT 4WH101, CTL K6, SCR 8-19-26

795  BLT 1900, REB 1910, CR T 1, MOT 4WH307F, CTL K35G, SCR 7-12-29

796  BLT 1899, REB 1910, CR T 2A, MOT 4WH101, CTL K6, SCR 7-6-26

797  BLT 1900, REB 1910, CR T 1, MOT 4WH307F, CTL K35G, SCR 3-28-30

798  BLT 1899, REB 1910, CR T 2A, MOT 4WH101, CTL K6, SCR 8-19-26

799  BLT 1900, REB 1910, CR T 1, car equipped with maximum traction TRKS, MOT 4WH307F, CTL K35G, SCR 8-28-30

### Fleet Statistics: Railroad Roof, 815–899

Roof: railroad; LEN: 815–817, 41' 8" 28' box; 818–899 43' 8" 30' box; W: 815–817, 39,400 lbs; 818–899, 44,600 lbs; seats: 38; crew size: 2-man; MOT: 815–817, 2WH101; 818–899, 4WH307F; FEN: Eclipse; HDL: Dayton 348A; types: 815–817, 700-type cars; 818–899, 800-type; TRKS: Brill 27F. All cars CCR except where noted.

815  10 windows, front and rear platform, DR, CR T 3B, became a waiting room at Brooklyn Station in 1926, SCR ?

816  BLT CCR 1903, REB CR 1910, DR, CR T 3, CTL K12, SCR 7-6-26

817  BLT CER 1904, REB CR 1910, DR, CR T 3B, CTL K10, SCR 7-2-26

818  BLT Kuhlman 1902, FN 638, REB Kuhlman 1910, CR T 6, CTL K6, SCR 8-22-29

819  BLT CER 1905, REB CR in 1910, CR T 6D, CTL K35G, SCR 9-3-29

820  BLT Brill 1901, CR T 5, Damaged in 1924 ACC, SCR 1924

821  BLT Kuhlman 1902, FN 636, REB in 1910, CR T 6, CTL K6, SCR 9-13-29

822  BLT Brill 1901, CR T 7, SCR 1924

823  BLT Brill 1901, CR T 7, SCR 1924

824  BLT Brill 1900, CR T 5, wrecked by train on Kinsman 2-5-11, SCR 1911

825  BLT Brill 1902, CR T 5, hit by freight train on Kinsman 10-12-20, SCR 1924

Car 893, a railroad roof 800, Brill, 1901. *(Cleveland Railway photo, Blaine Hays Collection)*

| | |
|---|---|
| 826 | BLT Brill 1902, FN 608, CR T 5, REB 1910, CTL K6, SCR 10-4-29 |
| 827 | BLT Brill 1901, CR T 7, SCR 1924 |
| 828 | BLT Brill 1900, FN 488, REB 1910, CR T 5, CTL K6, SCR 7-3-29 |
| 829 | BLT Brill 1901, FN 572, REB 1911, CR T 6C, CTL K35G, SCR 10-11-29 |
| 830 | BLT Brill 1901, FN 520, REB 1911, CR T 6C, CTL K35G, SCR 12-2-31 |
| 831 | BLT Brill 1900, CR T 7, SCR 1924 |
| 832 | BLT Brill 1902, FN 650, REB 1911, CR T 6C, W 43,600 lbs, CTL K35G, SCR 8-17-31 |
| 833 | BLT Brill 1900, CR T 6C, SCR 1924 |
| 834 | BLT Brill 1899, CR T 5, SCR 1924 |
| 835 | BLT Brill 1901, FN 562, REB 1910, CR T 6, CTL K6, SCR 11-5-31 |
| 836 | BLT Brill 1901, FN 552, REB 1911, CR T 5, CTL K6, SCR 10-11-29 |
| 837 | BLT Brill 1900, CR T 5A, SCR 1924 |

| | |
|---|---|
| 838 | BLT Brill 1902, FN 676, REB 1910, CR T 5, CTL K6, SCR 11-11-31 |
| 839 | BLT Kuhlman 1902, FN 634, REB 1910, CR T 5A, CTL K35G, SCR 12-2-31 |
| 840 | BLT Brill 1902, FN 648, REB 1910, CR T 5, CTL K6, SCR 6-25-31 |
| 841 | BLT Kuhlman 1901, FN 566, REB 1910, CR T 5A, CTL K35G, SCR 11-23-31 |
| 842 | BLT Brill 1900, FN 436, REB 1911, CR T 5A, CTL K35G, SCR 12-2-31 |
| 843 | BLT Brill 1902, FN 624, REB 1911, CR T 5A, CTL K35G, SCR 11-9-31 |
| 844 | BLT Brill 1902, CR T 5, SCR 1924 |
| 845 | BLT Brill 1901, FN 524, REB 1911, CR T 5A, CTL K35G, SCR 9-21-31 |
| 846 | BLT Brill 1902, REB 1910, CR T 5, CTL K6, SCR 11-9-31 |
| 847 | BLT Brill 1901, FN 536, REB 1910, CR T 5A, CTL K35G, SCR 11-11-31 |
| 848 | BLT Brill 1900, REB 1910, FN 486, CR T 5A, CTL K35G, SCR 12-2-31 |
| 849 | BLT Brill 1900, FN 478, REB 1910, CR T 5, CTL K6, SCR 11-17-31 |
| 850 | BLT Kuhlman 1902, FN 662, REB 1910, CR T 5, CTL K6, SCR 11-13-31 |
| 851 | BLT Brill 1902, FN 636, REB 1910, CR T 5A, CTL K35G, SCR 11-17-31 |
| 852 | BLT Brill 1901, FN 558, REB 1910, CR T 5A, CTL K35G, SCR 11-23-31 |
| 853 | BLT Brill 1900, FN 438, REB 1911, CR T 5A, CTL K35G, SCR 11-25-31 |
| 854 | BLT Brill 1901, REB 1911, CR T 5A, CTL K35G, SCR 11-9-31 |
| 855 | BLT Brill 1900, FN 468, REB 1910, CR T 5, CTL K6, SCR 11-19-31 |
| 856 | BLT Brill 1902, FN 606, REB 1910, CR T 5A, CTL K35G, SCR 11-19-31 |
| 857 | BLT Brill 1901, FN 516, REB 1910, CR T 5, CTL K6, SCR 9-17-31 |
| 858 | BLT Brill 1901, FN 556, REB 1910, CR T 5, CTL K6, SCR 7-13-31 |
| 859 | BLT Brill 1902, FN 622, REB 1910, CR T 5A, CTL K35G, SCR 9-23-31 |
| 860 | BLT Brill 1900, FN 476, REB 1910, CR T 5, CTL K6, SCR 7-13-31 |
| 861 | BLT Brill 1900, FN 614, REB 1911, CR T 5, CTL K6, SCR 7-27-31 |

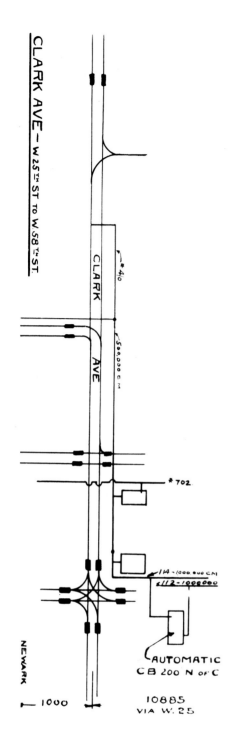

| | |
|---|---|
| 862 | BLT Brill 1901, FN 502, REB 1910, CR T 5A, CTL K35G, SCR 7-21-31 |
| 863 | BLT Jewett 1900, FN 418, REB 1910, CR T 5A, CTL K35G, SCR 8-3-31 |
| 864 | BLT Brill 1899, CR T 5, SCR 1924 |
| 865 | BLT Brill 1901, CR T 5, SCR 1924 |
| 866 | BLT Brill 1901, FN 546, REB 1911, CR T 5A, CTL K35G, SCR 11-25-31 |
| 868 | BLT Jewett 1900, CR T 5, SCR 1924 |
| 869 | BLT Brill 1904 for CER, CR T 5, SCR 1924 |
| 870 | BLT Brill 1900, CR T 5, SCR 1924 |
| 871 | BLT Brill 1902, FN 628, REB 1910, CR T 5, CTL K6, SCR 9-28-29 |
| 872 | BLT Kuhlman 1899, FN 392, REB 1911, CR T 5A, CTL K35G, SCR 9-20-29 |
| 873 | BLT Brill 1901, REB 1910, CR T 5, CTL K6, SCR 10-22-31 |
| 874 | BLT Brill 1901, FN 542, REB 1911, CR T 5, CTL K6, SCR 8-5-31 |
| 875 | BLT Kuhlman 1900, FN 474, REB 1910, CR T 5, CTL K6, SCR 7-23-31 |
| 876 | BLT Brill 1900, FN 464, REB 1911, CR T 5, CTL K6, SCR 11-17-31 |
| 877 | BLT Kuhlman 1902, FN 642, REB 1910, CR T 5A, CTL K35G, SCR 11-13-31 |
| 878 | BLT Brill 1902, FN 610, REB 1910, CR T 5A, CTL K35G, SCR 11-19-31 |
| 879 | BLT Kuhlman 1901, FN 498, REB 1910, CR T 5, CTL K6, SCR 11-11-31 |
| 880 | BLT Brill 1902, FN 620, REB 1911, CR T 5A, CTL K35G, SCR 11-13-31 |
| 881 | BLT Brill 1902, FN 678, REB 1911, CR T 5A, CTL K35G, SCR 7-29-31 |
| 882 | BLT Brill 1899, FN 402, REB 1910, CR T 5A, CTL K35G, SCR 7-29-31 |
| 883 | BLT Brill 1901, FN 528, REB 1910, CR T 5, CTL K6, SCR 9-25-31 |
| 884 | BLT Brill 1899, FN 412, REB 1910, CR T 5, CTL K6, SCR 8-5-31 |
| 885 | BLT Brill 1901, FN 534, REB 1910, CR T 5A, CTL K35G, SCR 7-3-31 |
| 886 | BLT Brill 1901, FN 492, REB 1910, CR T 5A, CTL K35G, SCR 7-15-29 |
| 887 | BLT Brill 1902, FN 626, REB 1910, CR T 5, CTL K6 car damaged in collision at Harvard and Broadway 4-5-22, SCR 9-18-29 |
| 888 | BLT Brill 1901, REB 1911, CR T 5A, CTL K35G, SCR 9-6-29 |

889  BLT Brill 1900, CR T 5, SCR 1924
890  BLT Brill 1901, FN 560, REB 1910, CR T 5, CTL K6, SCR 9-11-31
891  BLT Brill 1901, FN 574, REB 1910, CR T 5, CTL K6, SCR 9-24-29
892  BLT Brill 1901, FN 500, REB 1910, CR T 5A, CTL K35G, SCR 8-26-29
893  BLT Brill 1901, FN 530, REB 1911, CR T 5A, CTL K35G, SCR 10-3-29
894  BLT Brill 1901, FN 569, REB 1910, CR T 5, CTL K6, SCR 10-4-29
895  BLT Brill 1901, FN 528, REB 1911, CR T 5A, CTL K35G, SCR 9-16-29
896  BLT Brill 1901, FN 580, REB 1910, CR T 5, CTL K6, SCR 10-1-29
897  BLT Brill 1901, FN 494, REB 1910, CR T 5, CTL K6, SCR 9-26-29
898  BLT probably by Brill in 1901?, CR T 5, on 1-5-21 car had CTL problems and slid down Cedar Hill, burned, SCR 1921
899  BLT Brill 1901, FN 544, REB 1910, CR T 5A, wrecked on Union 12-22-20, CTL K35G, SCR 10-1-29

### Fleet Statistics: Deck Roof, 900–999

BLDR: 900, Cincinnati 1908; 901–919 and 921–925, Kuhlman 1910; 920 (FN 415) and 926–949, Stephenson 1906; 950–999, St. Louis 1907; LEN: 901–925, 52' 6"; 926–999, 45'; (900–925 REB to Peter Witt configuration beginning in 1915); FEN: Eclipse (all Stephenson-built cars CON to SE in 1908); CTL: K28B; style: 900–925, box; 926–999, semi-convertible; W: 900–925, 48,480 lbs; 926–999, 49,900 lbs; crew size: two; HDL: Dayton 348A; seats: 55; CR T: 900–925, T 18; 926–943, T 9; 944–949, T 10; 950–999, T 11; TRKS: 900–949, Brill 27G-1; 950–999, Brill 27F; MOT: 900–999, 4WH101.

900  BLT 1908, LEN 52', MOT 4WH307F, SCR 7-29-31
901  REB 2-10-15, MOT 4WH 307F, SCR 8-1-29
902  D 2-1910, MOT 4WH307F, SCR 8-10-31
903  REB 2-10-15, MOT 4WH307F, SCR 8-26-29
904  REB 3-10-15, MOT 4WH307F, SCR 8-19-29
905  REB 2-10-15, MOT 4WH307F, SCR 9-16-29
906  BLT 2-10-10, MOT 4WH307F, SCR 8-3-31
907  REB 1915, MOT 4WH307F, SCR 9-20-29
908  BLT 3-1910, ACC on Superior 12-22-20, MOT 4WH307F, SCR 9-26-29
909  BLT 3-1910, REB 1915, SCR 9-28-29

Car 987, FN 3387, former Tom Johnson Car, St. Louis, 1907, CR type 11. *(Cleveland Railway photo, Blaine Hays Collection)*

| | |
|---|---|
| 910 | BLT 3-1910, SCR 8-5-31 |
| 911 | BLT 3-1910, SCR 7-27-31 |
| 912 | BLT 3-1910, SCR 9-24-29 |
| 913 | BLT 3-1910, SCR 7-21-31 |
| 914 | REB 3-10-15, SCR 9-18-29 |
| 915 | BLT 3-1910, SCR 7-13-31 |
| 916 | BLT 3-1910, SCR 7-23-31 |
| 917 | BLT 3-1910, SCR 7-16-31 |
| 918 | BLT 3-1910, SCR 10-4-29 |
| 919 | BLT 3-1910, SCR 6-25-31 |
| 920 | BLT 3-1910, REB 1915, TRKS B27F, CTL K6, destroyed in wreck St. Clair–East 79th 10-1-18 |

920(II)  BLT 1899, FN 415, refurbished in 1919 to replace first 920, only curved sided Peter Witt car ever built, SCR 7-17-26
921  BLT 3-1910, SCR 9-20-29
922  BLT 3-1910, SCR 7-1-31
923  BLT 3-1910, SCR 7-8-31
924  BLT 3-19-10, SCR 7-3-31
925  BLT 3-1910, SCR 7-15-29
926  BLT ?, FN 3326, REB 1913, SCR ?
927  BLT ?, FN 3327, REB 1913, SCR ?
928  BLT 1906, FN 3328, REB 1913, MOT 4GE80, SCR 7-14-26
929  BLT ?, FN 3329, REB 1913, SCR 1925
930  BLT ?, FN 3330, REB 1913, SCR 1925
931  BLT 1906, FN 3331, REB 1913, MOT 4GE80, SCR 7-15-26
932  BLT 1906?, FN 3332, SCR 1925
933  BLT 1906?, FN 3333, REB 1913, SCR 1925
934  BLT 1906?, FN 3334, REB 1913, sold in 1924 to Nelson, B.C. and REN 23, retired to storage status and restored in the 1980s and presently operates on a one-mile line near downtown Nelson.
935  BLT 1906, FN 3335, REB 1913, SCR 7-17-26
936  BLT 1900, FN 3336, REB 1913, SCR 7-22-26
937  BLT 1906?, FN 3337, REB 1913, SCR 1925
938  BLT 1906, FN 3338, REB 1913, MOT 4GE80, SCR 7-15-26
939  BLT 1906?, FN 3339, REB 1913, SCR 1925
940  FN 3340, REB 1913, MOT 4GE80, SCR 7-14-26
941  FN 3341, REB 1913, SCR 7-20-26
942  FN 3342, REB 1913, SCR 7-20-26
943  FN 3343, REB 1913, SCR 7-24-26
944  FN 3344, REB 1913, SCR 11-14-25
945  FN 3345, REB 1913, SCR 1925
946  FN 3346, REB 1913, SCR 1925
947  FN 3347, REB 1913, SCR 1925
948  FN 3348, REB 1913, SCR 1925
949  FN 3349, REB 1913, MOT 4WH101B, SCR 7-22-26
950  FN 3350, REB 1913, SCR 1925
951  FN 3351, REB 1913, SCR 1925
952  FN 3352, REB 1913, SCR 1925
953  FN 3353, REB 1913, CR T9A, SCR 1925
954  FN 3354, REB 1913, MOT 4WH101B-6, TRKS B27F-E-1, SCR 8-4-27
955  FN 3355, REB 1913, CR T9A, SCR 6-28-26
956  FN 3356, REB 1913, SCR 8-7-28
957  FN 3357, REB 1913, MOT 4WH101B-6, TRKS B27F-E-1, SCR 1927
958  FN 3358, REB 1913, MOT 4WH101B-6, TRKS B27F-E-1, SCR 7-24-28

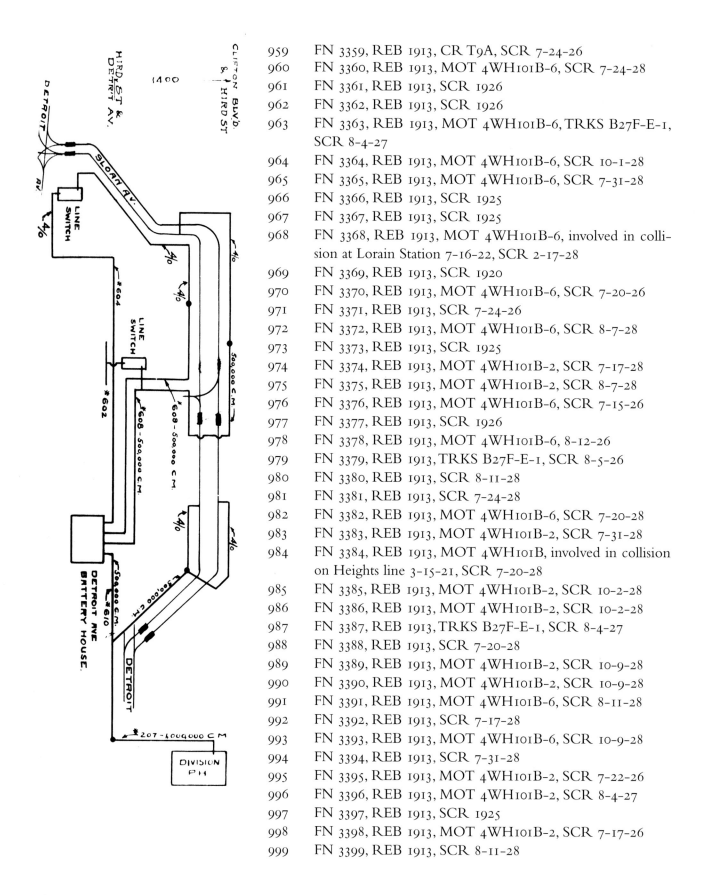

| | |
|---|---|
| 959 | FN 3359, REB 1913, CR T9A, SCR 7-24-26 |
| 960 | FN 3360, REB 1913, MOT 4WH101B-6, SCR 7-24-28 |
| 961 | FN 3361, REB 1913, SCR 1926 |
| 962 | FN 3362, REB 1913, SCR 1926 |
| 963 | FN 3363, REB 1913, MOT 4WH101B-6, TRKS B27F-E-1, SCR 8-4-27 |
| 964 | FN 3364, REB 1913, MOT 4WH101B-6, SCR 10-1-28 |
| 965 | FN 3365, REB 1913, MOT 4WH101B-6, SCR 7-31-28 |
| 966 | FN 3366, REB 1913, SCR 1925 |
| 967 | FN 3367, REB 1913, SCR 1925 |
| 968 | FN 3368, REB 1913, MOT 4WH101B-6, involved in collision at Lorain Station 7-16-22, SCR 2-17-28 |
| 969 | FN 3369, REB 1913, SCR 1920 |
| 970 | FN 3370, REB 1913, MOT 4WH101B-6, SCR 7-20-26 |
| 971 | FN 3371, REB 1913, SCR 7-24-26 |
| 972 | FN 3372, REB 1913, MOT 4WH101B-6, SCR 8-7-28 |
| 973 | FN 3373, REB 1913, SCR 1925 |
| 974 | FN 3374, REB 1913, MOT 4WH101B-2, SCR 7-17-28 |
| 975 | FN 3375, REB 1913, MOT 4WH101B-2, SCR 8-7-28 |
| 976 | FN 3376, REB 1913, MOT 4WH101B-6, SCR 7-15-26 |
| 977 | FN 3377, REB 1913, SCR 1926 |
| 978 | FN 3378, REB 1913, MOT 4WH101B-6, 8-12-26 |
| 979 | FN 3379, REB 1913, TRKS B27F-E-1, SCR 8-5-26 |
| 980 | FN 3380, REB 1913, SCR 8-11-28 |
| 981 | FN 3381, REB 1913, SCR 7-24-28 |
| 982 | FN 3382, REB 1913, MOT 4WH101B-6, SCR 7-20-28 |
| 983 | FN 3383, REB 1913, MOT 4WH101B-2, SCR 7-31-28 |
| 984 | FN 3384, REB 1913, MOT 4WH101B, involved in collision on Heights line 3-15-21, SCR 7-20-28 |
| 985 | FN 3385, REB 1913, MOT 4WH101B-2, SCR 10-2-28 |
| 986 | FN 3386, REB 1913, MOT 4WH101B-2, SCR 10-2-28 |
| 987 | FN 3387, REB 1913, TRKS B27F-E-1, SCR 8-4-27 |
| 988 | FN 3388, REB 1913, SCR 7-20-28 |
| 989 | FN 3389, REB 1913, MOT 4WH101B-2, SCR 10-9-28 |
| 990 | FN 3390, REB 1913, MOT 4WH101B-2, SCR 10-9-28 |
| 991 | FN 3391, REB 1913, MOT 4WH101B-6, SCR 8-11-28 |
| 992 | FN 3392, REB 1913, SCR 7-17-28 |
| 993 | FN 3393, REB 1913, MOT 4WH101B-6, SCR 10-9-28 |
| 994 | FN 3394, REB 1913, SCR 7-31-28 |
| 995 | FN 3395, REB 1913, MOT 4WH101B-2, SCR 7-22-26 |
| 996 | FN 3396, REB 1913, MOT 4WH101B-2, SCR 8-4-27 |
| 997 | FN 3397, REB 1913, SCR 1925 |
| 998 | FN 3398, REB 1913, MOT 4WH101B-2, SCR 7-17-26 |
| 999 | FN 3399, REB 1913, SCR 8-11-28 |

Car 907, deck-roof type, 900 series as rebuilt, G. C. Kuhlman, 1910, CR type 18. *(Cleveland Railway photo, Blaine Hays Collection)*

Car 920 II, FN 415, the only curve-side Peter Witt Car ever built. *(Cleveland Railway photo, Blaine Hays Collection)*

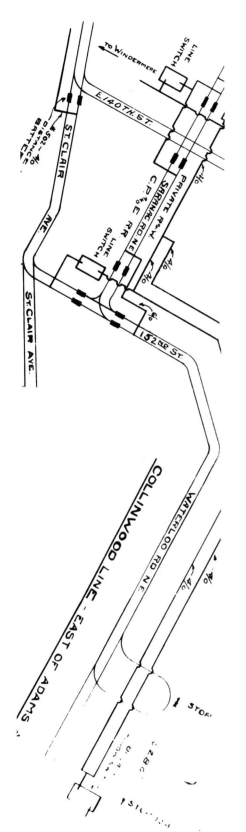

## Fleet Statistics: Dinkeys

LEN: 43' 3"; W: 44,600 lbs; seating: 38; TRKS: Brill 27F; MOT: Westinghouse #101; HDL: Dayton 348A and Golden Glow; FEN: Eclipse.

| | |
|---|---|
| 42 | FN 387, BLT Kuhlman 1902, two K35G CTL, B27F TRKS, Eclipse wheel guards, dead storage at Brooklyn 1939, SCR 4-3-41. |
| 43 | FN 338, BLT Brill 1903, two K35G CTL, B27F TRKS, wrecked and stored at Brooklyn. Christiansen roster says FN was 388; Harvard Shop record says 338 but is listed as SCR 1940. Car 338 became trailer and sold to Northern Ohio Traction 1918, SCR 3-20-40 |
| 44 | FN 389, BLT Kuhlman 1902, two K35G CTL, B27F TRKS, ran from E. 55th, SCR 3-6-40 |
| 45 | FN 391, BLT Kuhlman 1902, two K-12 CTL, B27F TRKS, ran from Superior, dead storage 1939, SCR 4-11-41 |
| 46 | FN 393, BLT Kuhlman in 1902, two K35G CTL, B27F TRKS, Eclipse special fenders, operated from E. 55th, wrecked and stored, SCR 3-18-40 |
| 47 | FN 395, BLT Kuhlman in 1902, two K35G CTL, B27F TRKS, dead storage at E. 55th 1939, SCR 4-9-41 |
| 48 | FN 397, BLT Kuhlman in 1902, two K35G CTL, B27F TRKS, dead storage at E. 55th 1939, SCR 4-9-41 |
| 49 | FN 399, BLT Kuhlman in 1904, two K35G CTL, B27F TRKS, Eclipse wheel guards, ran from E. 55th, wrecked, SCR 3-19-40 |
| 111 | BLT Niles 1903, two K10 CTL, B27F TRKS, Eclipse wheel guards, W 39,400 lbs., SCR 1-21-36 |
| 188 | BLT Niles 1903, two 35G CTL, B27F TRKS, W 39,400 lbs., Eclipse wheel guards, SCR 1-21-36 |
| 193 | BLT Niles 1903, two K12 CTL, B27F TRKS, W 39,400 lbs., Eclipse wheel guards, ACC, operated out of Brooklyn, in dead storage 1939, SCR 3-21-41 |
| 197 | BLT Niles 1903, CON to DE dinkey 1921, CON to rail bond test car 0151 8-11-28, SCR 12-6-51 |
| 800 | BLT Niles in 1903, REB 1911, REB to DE dinkey in 1917, CR T 3E, FEN Eclipse wheel guards, HDL 1 Golden Glow & 1 Dayton, CTL 2 K12, ACC 2-26-21, five killed in auto on Abbey, SCR E. 34th Station 1-20-36 |
| 801 | BLT Niles 1903, REB 1911, REB to DE dinkey in 1917, CR T 3A, FEN eclipse wheel guards, HDL 1 Golden Glow and 1 Dayton, CTL 2 K10, SCR 12-8-31 |
| 802 | BLT Niles in 1903, REB 1911, REB to DE dinkey in 1917, CR T?, FEN 2 eclipse wheel guards, HDL 1 Golden Glow |

Car 47, railroad-roof dinkey type, G. C. Kuhlman, 1902. *(Cleveland Railway photo, Blaine Hays Collection)*

and 1 Dayton, CTL 2 K12, TRKS B27F-E-1, involved in collision at Mayfield and Taylor 7-17-22, later assigned to Brooklyn Station, SCR E. 34th Station 1-21-36

803  BLT Niles 1903, REB to DE dinkey in 1911, CR T 3A, FEN 2 Eclipse wheel guards, HDL 1 Golden Glow and 1 Dayton, CTL 2 K10, used as an office at Harvard yard, dead storage at Miles Station in 1939, underframe, SCR Harvard 4-16-41, body burned 7-1-42

804  BLT Niles 1903, REB to DE dinkey in 1911, CR T 3A, FEN 2 Eclipse wheel guards, HDL 1 Golden Glow and 1 Dayton, CTL 2 K10, MTR 2WH101B, became money box storage car at East 55th yard, SCR 7-10-35

805  BLT Niles 1903, REB to DE dinkey 1911, CR T 3A, FEN 2 Eclipse wheel guards, HDL 1 Golden Glow and 1 Dayton, CTL 2 K12, damage from derailment E. 49th Street 9-21-21, stored Miles Station, SCR E. 34th Station 1-11-36

806  BLT Niles 1903, REB to DE dinkey in 1911, CR T 3A, FEN 2 Eclipse wheel guards, HDL 1 Golden Glow and 1 Dayton,

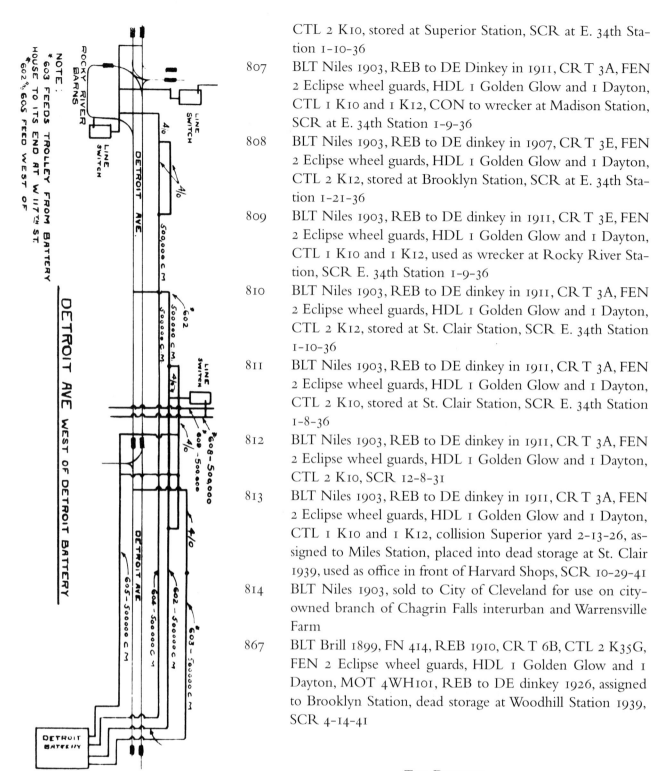

|     | |
| --- | --- |
|     | CTL 2 K10, stored at Superior Station, SCR at E. 34th Station 1-10-36 |
| 807 | BLT Niles 1903, REB to DE Dinkey in 1911, CR T 3A, FEN 2 Eclipse wheel guards, HDL 1 Golden Glow and 1 Dayton, CTL 1 K10 and 1 K12, CON to wrecker at Madison Station, SCR at E. 34th Station 1-9-36 |
| 808 | BLT Niles 1903, REB to DE dinkey in 1907, CR T 3E, FEN 2 Eclipse wheel guards, HDL 1 Golden Glow and 1 Dayton, CTL 2 K12, stored at Brooklyn Station, SCR at E. 34th Station 1-21-36 |
| 809 | BLT Niles 1903, REB to DE dinkey in 1911, CR T 3E, FEN 2 Eclipse wheel guards, HDL 1 Golden Glow and 1 Dayton, CTL 1 K10 and 1 K12, used as wrecker at Rocky River Station, SCR E. 34th Station 1-9-36 |
| 810 | BLT Niles 1903, REB to DE dinkey in 1911, CR T 3A, FEN 2 Eclipse wheel guards, HDL 1 Golden Glow and 1 Dayton, CTL 2 K12, stored at St. Clair Station, SCR E. 34th Station 1-10-36 |
| 811 | BLT Niles 1903, REB to DE dinkey in 1911, CR T 3A, FEN 2 Eclipse wheel guards, HDL 1 Golden Glow and 1 Dayton, CTL 2 K10, stored at St. Clair Station, SCR E. 34th Station 1-8-36 |
| 812 | BLT Niles 1903, REB to DE dinkey in 1911, CR T 3A, FEN 2 Eclipse wheel guards, HDL 1 Golden Glow and 1 Dayton, CTL 2 K10, SCR 12-8-31 |
| 813 | BLT Niles 1903, REB to DE dinkey in 1911, CR T 3A, FEN 2 Eclipse wheel guards, HDL 1 Golden Glow and 1 Dayton, CTL 1 K10 and 1 K12, collision Superior yard 2-13-26, assigned to Miles Station, placed into dead storage at St. Clair 1939, used as office in front of Harvard Shops, SCR 10-29-41 |
| 814 | BLT Niles 1903, sold to City of Cleveland for use on city-owned branch of Chagrin Falls interurban and Warrensville Farm |
| 867 | BLT Brill 1899, FN 414, REB 1910, CR T 6B, CTL 2 K35G, FEN 2 Eclipse wheel guards, HDL 1 Golden Glow and 1 Dayton, MOT 4WH101, REB to DE dinkey 1926, assigned to Brooklyn Station, dead storage at Woodhill Station 1939, SCR 4-14-41 |

## The Dinkeys
*by Bob Korach*

The strangest of Cleveland streetcars were the twenty-eight dinkey cars that ran in revenue service from 1907 to 1939. They were double-ended wooden cars rebuilt from single-end cars and used on lines

without loops or wyes. Many of these operations substituted for or were replaced by regular operations, but others were purely dinkey operations. Cleveland Railway inherited seven of these cars in 1910 and added twenty-one of its own up to 1926.

The first of the dinkeys to go was 814, which was sold to the City of Cleveland in 1914 for use on the city-owned branch of the Chagrin Falls interurban at the Warrensville Farm off Chagrin Boulevard where it ran along with car 2. The line ceased operating in 1925, and the cars were scrapped.

The Abbey line originally switched back at West 14th–Abbey on the east end and West 25th–Lorain on the west. This end was cut back to Abbey and Lorain when the Lorain-Carnegie Bridge opened around 1934. The Superior line only ran a few trips using car 45 on weekdays for school riding but ran all Sunday morning for regular riders. Car 803 was the spare car at Superior Station.

The Clark Bridge line used at least four cars in the peak periods for steel mills and even had conductors in rush hours. The State Road line was single track with a passing siding in the center of the line. In spite of this, on October 18, 1937, cars 193 and 43 met head on in a

Car 807, railroad-roof type, Niles, 1903, CR type 3A. *(Cleveland Railway photo, Blaine Hays Collection)*

serious collision. And soon thereafter cars 46 and 49 collided as well. Thus, four dinkeys were quietly retired. The final cars on State were 42, 44, 47, 48, and 867, with 47 being the last regular car and 867 the last ceremonial car. Cars 45 and 803 remained at Superior Station, and at East 55th 813 stood ready for a call that never came.

### Fleet Statistics: Funeral Cars

LEN: old, 42'; new, 44'6"; MOT: old, 4WH#49; new, 4WH#101; HDL: old, Dash; new, Imperial Arc; FEN: Eclipse.

*Four Casket (old):* REB by CER at Lakeview Shop in 1898, had K6 CTL, B27F TRKS; was a combine (baggage compartment) purchased used from Akron, Bedford and Cleveland Railroad; car had no number; painted black on 10-13-18; carried an extra trolley pole; SCR 3-29-24.

*Two Casket (new):* BLT by CER in 1905, K6 CTL, P25M TRKS; BLT new atop Peckham TRKS at Lakeview Shops; painted "dead black," according to the record, on 5-14-18; operated from Cedar carhouse; body was sold sans equipment 11-18-24.

### Fleet Statistics: The Holmden Dinkey

2 (FN possibly 202), also could have been 501 or 553 (ST), known as the Holmden Dinkey and sold to the City of Cleveland for Warrensville farm service; SCR 9-8-14.

### Fleet Statistics: Experimental Subway Car

101, BLT Kuhlman 6-11-13, had a K12 CTL and B39E1 maximum-traction TRKS; car had sliding doors, seated 57, and WH307 MOT. The only passenger car on CR with a whistle (cars 4050 and 4051 carried whistles much later during operation on the CIRR). Car BLT for Cleveland Underground Subway Company as sample subway car. Motorman had high-platform front door, and passengers used drop-center double doors. Car loosely inspired by "stepless" Hedley-Doyle hobbleskirt cars of New York Railways. Subway proposal was for under Euclid Avenue, from Public Square to University Circle, but idea was not pursued and car reverted to CR. It was then "buried" at Lorain Station and operated as the Linndale "dinkey" until SCR on 7-3-29.

### Fleet Statistics: "800-Type" Cars (SE)

40   FN 332, BLT Brill, 1903, had K-6 CTL and B27F TRKS, SCR 7-8-31

Car 101, experimental Cleveland Underground subway Car, G. C. Kuhlman, 1912. *(Eugene Schmidt Collection, Northern Ohio Railway Museum)*

41  FN 342, BLT Brill, 1903, had K-6 CTL and B27E TRKS, SCR 10-11-29

99  FN 324, (ST), BLT CER, 1893, two K-10 CTL and Dupont TRK, car was CER Parlor Car, became sightseeing car and CON to welder 024, CON to waiting shelter at St. Clair and East 222 in 1922, SCR 11-3-26

115  T 6C, BLT Brill 1903, K6 CTL, B27F TRKS, SCR 7-1-31
118  T 6C, BLT Brill 1903, K6 CTL, B27F TRKS, SCR 7-21-31
119  T 6C, BLT Brill 1903, K6 CTL, B27F TRKS, SCR 7-3-31
121  T 6C, BLT Brill 1903, K6 CTL, B27F TRKS, SCR 8-10-31
122  T 6C, BLT Brill 1903, K6 CTL, B27F TRKS, SCR 7-21-31
123  T 6C, BLT Brill 1903, K6 CTL, B27F TRKS, SCR 6-25-31
191  BLT Brill 1903, K35G CTL, B27F TRKS, SCR 7-16-31
192  T 4A, BLT Brill 1903, K6 CTL, B27F TRKS, SCR 10-1-31

## SECTION 4
### Passenger Equipment Purchased by Cleveland Railway and Cleveland Transit

On March 1, 1910, the new Cleveland Railway Company took over operations of all Cleveland Electric Railway and Municipal Traction operations under a court-ordered service-at-cost formula known as the Tayler Franchise, named after federal judge Robert Walker Tayler who proposed this compromise solution to street railway problems.

The most urgent need was for new equipment to meet ever-growing ridership demands. Immediately upon takeover, the new company placed an order for twenty-five tremendously long fifty-two-foot cars based on Tom L. Johnson's number 900. This set the stage for the era of big cars, and the large order of one hundred trailer cars three years later began the large trailer era.

The development of a front-entrance, center-exit car in 1914 by traction commissioner Peter Witt introduced the motor-trailer era for which Cleveland became famous. Following is a count of the 1,432 passenger cars of Cleveland Railway and Cleveland Transit. Until the second decade, work cars were built especially for a given use. With so many passenger cars coming out of service from the old systems, many cars were tapped by Cleveland Railway to serve as work units. The following list is in order of acquisition.

## Outline of Cleveland Railway/Transit Streetcars

| Number of Cars | Year Built | Builder | System Built for | Order/Car Series | Car Notes |
|---|---|---|---|---|---|
| 25 | 1910 | Kuhlman | CR | 901–925 | DR, 14 windows |
| 1 | 1911 | CR | CR | 1052 (FN 824) | Scullin louver roof, 14 windows |
| 1 | 1912 | Kuhlman | Cleveland Underground Subway Co. | 101 | low platform center, high platform front |
| 52 | 1913 | Kuhlman | CR | 1000–1051 | louver roof, 14 window |
| 100 | 1913 | Kuhlman | CR | 2000–2099 | CE trailers |
| 200 | 1913 | Kuhlman | CR | 2100–2299 | CE trailers |
| 201 | 1914–15 | Kuhlman | CR | 1100–1300 | louver roof, CE cars |
| 1 | 1914 | CR | CR | 330 (FN 33) | first Peter Witt |
| 50 | 1915 | Kuhlman | CR | 200–249 | Peter Witt |
| 80 | 1916 | Kuhlman | CR | 250–329 | Peter Witt |
| 26 | 1917 | Kuhlman | CR | 2300–2325 | CE trailers |
| 25 | 1917 | CR | CR | 2326–2350 | CE trailers |
| 25 | 1918 | Kuhlman | CR | 2351–2375 | CE trailers |
| 25 | 1918 | Cincinnati | Rochester | 1075–1099 | "Submarine" Peter Witts |
| 50 | 1920 | Kuhlman | CR | 2451–2500 | CE trailers |
| 50 | 1920 | Kuhlman | CR | 450–499 | Peter Witt |
| 50 | 1923 | Kuhlman | CR | 2376–2425 | CE trailers |
| 100 | 1923 | Kuhlman | CR | 350–449 | Peter Witt |
| 25 | 1924–25 | CR | CR | 150 (FN 1301)–174 | Peter Witt |
| 16 | 1924 | CR | CR | 1982–1997 | CE trailers |
| 75 | 1924–25 | CR | CR | 1301–1375 | Peter Witt |
| 1 | 1926 | CR | CR | 1376 | all-aluminum Peter Witt |
| 28 | 1927–29 | Kuhlman | CR | 5000–5027 | articulated Peter Witt |
| 150 | 1928–30 | Kuhlman | CR | 4000 (FN 333)–4149 | Peter Witt |
| 50 | 1946 | Pullman | CTS | 4200–4249 | PCC streetcar |
| 25 | 1946 | St. Louis | CTS | 4250–4274 | PCC streetcar |

## NOTES

1. Trailers 2200 to 2220 leased to Detroit, MI, 1928–29. Cars 747, 749, 751, 753, 755, 757, 761, 763, 765, 767, 769, 771, 773, 779, 781, 783, 787, 791, 793, 795, 799 were also leased and pulled the trailers; were delivered over the Lake Shore Electric and Eastern Michigan interurban lines.
2. Cars 901 to 925 were ordered by CER and D after takeover to CR; became the only wooden cars built for CR.
3. Twenty-five 1918-built Peter Witt "Submarine" cars were built for Rochester, NY, by Cincinnati Car Co. but were diverted to Cleveland. They ran less than five years and were sold as follows:

   1922: 1085–1086, 1092–1093, 1096 to Kitchener, Ontario, and SCR there in 1947
   
   1923: 1075, 1079, 1081–1082, 1091 to London, Ontario, and from there resold to Saskatoon, Saskatchewan, and REN 201–205.
   
   1924: 1076, 1077, 1083, 1084, 1095 to Lake Shore Electric's Lorain Street Railroad and REN 80–84; resold to Indianapolis in 1938 and REN 1057–1061.

Car 1080, Submarine Peter Witt, Cincinnati, 1918. *(Cleveland Railway photo, Blaine Hays Collection)*

1924: 1078, 1080, 1087, 1088–1090, 1094, 1097–1099 to Indianapolis and REN 1034–1043; SCR after World War II.

4. Trailers 1982–1997 were built to match the 1301–1375 series.
Trailers 2300–2375 were built to match the 200–329 series.
Trailers 2376–2425 were built to match the 350–449 series.
Trailers 2451–2500 were built to match the 450–499 series.

5. Starting in 1939, a different paint design was developed by New York artist Raymond Loewy, famous for his vehicle designs. He changed the colors from plain yellow with red striping to a two-tone green and gray combination set in a modern scheme that featured a breakup of straight lines. Cars painted in this format are designated "G" in the roster. Problems with the new scheme arose, however: the green paint faded badly, dust on the cars was more apparent, and an increase in night accidents occurred because the vehicles could not be easily seen in the dark. So early in 1941 the colors (not the pattern) were changed to the two-tone mustard and cream with mahogany striping. These colors lasted until the introduction of dark and powder blues in the 1950s.

### One-Man Operation
*by Bob Korach*

John J. Stanley, president of the Cleveland Railway Company from 1910 until his death in 1926, always believed that if a transit line was worth having streetcars then it should have big streetcars, even if that required a motorman and conductor on each car.

All new streetcars bought after 1910 were over 50 feet long and among the largest in the trolley business. No little Birney cars were ever bought for passenger service, but Stanley did perceive that he could get some of the benefits of one-man operation by buying trailer cars requiring only a conductor. Over one-third of the fleet was thus one-man operated by 1912.

Little is known about whether the dinkeys that started in 1907 ever regularly used conductors, except that the Clark Bridge dinkeys were two-man operated in peak hours. The first feeble effort toward regular one-man operation was on April 1, 1920, when the almost-new but very lightly used East 30th Street line was converted to one-man operation using wooden Kuhlman cars in the low 500 series (500, 501, 503, 504, 510, 515, 517, 520, 521, 523, 527, 528), replacing steel cars 250–258. No real operating problems were encountered.

Six years passed before the next effort came about. Steel cars 200–207 were converted in 1926 to one-man cars to replace the wood cars on East 30th Street, and the wood cars went west to Clifton where they promptly failed the second one-man test.

When 4050–4149 were purchased in 1929 and 1930, they were partially designed to be one-man cars but went into service on the Mayfield, Fairmount, Superior, Euclid, Broadway, and Lorain lines as two-man cars, usually pulling trailers. In 1935, Heights cars 4050 and 4051 were equipped for full one-man operation and were assigned to the Shaker Rapid. Shaker soon decided to buy secondhand one-man cars, and 4050 and 4051 returned to the Heights lines and two-man operation.

The next effort at one-man operation was no more successful. In starting from scratch to make a suitable safety-equipped one-man car, 150–174 and 1311–1339 received extensive rebuilding in 1935. They went into service on January 1, 1936, on Harvard-Denison and Fulton, where they only survived three months due to passenger and labor union complaints. With the conversion of East 30th Street to bus on November 15, 1940, and the previous abandonment of the various dinkey lines, there was no one-man operation in Cleveland except for the seven one-man cars on the Shaker Rapid.

In 1941 buses replaced streetcars nights and Sundays on Woodland, Buckeye, Union, Broadway, and Clark so that the Cleveland Railway could have some of the benefits of one-man operation. During this same time period, Heights car 4068 was rebuilt into a very modern one-man car. It was taken out to the City of Lakewood one night and shown to the local officials as the proposed replacement for the by-then-ancient 1200-type two-man cars on Clifton. Officials there said thanks but no thanks, and the car returned to storage in the Harvard Shops. On February 1, 1942, wartime riding forced the return of 4068 to two-man service on the Heights line. This was the third failure at one-man operation.

In 1942 CTS was told by the federal government to put streetcars back on the lines using buses nights and Sundays. Thus another opportunity for one-man cars presented itself. The cars would run with conductors in the daytime and without conductors nights and Sundays. Previously converted cars 150–174 were assigned to Union; 1322–1339 were assigned to Woodland and Buckeye (those cars had backup controls) and had been previously used on Harvard-Denison in the 1936 experiment. Cars 1311–1321 plus seventeen of the 4050-type cars went to Broadway, and sixteen of the 4050 type to Clark (backup controls). With heavy wartime riding, this attempt soon came to an end, and 100 percent two-man cars returned.

Finally on October 1, 1946, twenty years after the first one-man failure on Clifton, PCC cars converted most of the Superior line and base service on St. Clair to one-man service. On May 1, 1947, the Pearl and State branches of the West 25th Street car line became PCC operated in all base and some peak service.

Using the same 150-, 1311-, and 4050-type cars, 100 percent one-man operation was introduced on the Kinsman line on September 7, 1947. Fifty-four cars were scheduled in the peak hour, replacing forty two-man cars. There was to be no failure this time! On July 1, 1948, came East 55th Street and Clark; Madison on June 11, 1949; Detroit on August 1, 1949; and Euclid and Lorain on March 25, 1950. Most of these lines were not 100 percent one-man operated at first, as two-man cars and even trailer trains were used in the peak hours.

Finally on June 14, 1952, after thirty-two years of trying to convert the Cleveland system to one-man operation, the last two-man car, 1343, made a round-trip on Lorain between the West 140th Street loop and Public Square, and the line became trolley bus the next day. The remaining lines, East 55th Street, Superior, Clark, West 25th Street, and Madison were all 100 percent one-man, but the streetcars themselves only lasted eighteen more months.

### CLEVELAND STREETCAR ROSTER OF EQUIPMENT (PASSENGER MOTOR CARS)

The base of information for this roster is the 3 x 5 Harvard Shop file car record cards rescued from RTA's Reed Garage by members of the Northern Ohio Railway Museum in 1983. Supplementing these are the CTS Inventory of Vehicles in two volumes, one for older vehicles, the other for buses and later vehicles. These were known to CTS/RTA employees as the "old and new" testaments and were gifts to one of the authors from RTA equipment administrator Charles K. Kocour.

In compiling this information, no supposition was accepted. Verification sources for roster information came from researchers Ken Morse and Harry Christiansen, both of whom published their rosters. When information was in conflict, the 3 x 5 cards and Inventory of Vehicles was accepted as the authority because this information was recorded closest to the time of the event and was the official corporate information. In cases where two verifiable records exist, both are listed and explained. Vehicle "delivery" dates (D) are the company's acceptance date, scrapping dates the day the cutting torch was first used. The date the equipment department requested a piece of equipment be scrapped and the actual scrapping date may vary greatly. Therefore no "request" dates were used in the roster. The Cleveland Railway and the Cleveland Transit System fleets are presented in order of acquisition.

In selecting roster photos, we have tried to include as many styles of equipment as possible. For economy, some vehicle equipment types appearing in the companion volume are not repeated here. In these instances there will be listings without accompanying photos.

# THE CLASSIC PERIOD
*by J. William Vigrass*

After the merger in 1910 that produced the Cleveland Railway Company, a classic period occurred in which streetcar development in Cleveland proceeded in a steady evolutionary pattern and culminated in a system that brought the art and science of hauling people down streets to a pinnacle that was seldom equaled and not surpassed during the period 1914–45.

Following the creation of Cleveland Railway under the service-at-cost Tayler Grant, it was immediately apparent that a substantial proportion of the existing car fleet was obsolete and due for replacement. The Tayler Grant had a significant impact on the technology selected by the railway. The low fare prescribed under the grant, initially three cents, drove the decision to have the largest car that would operate effectively on Cleveland's streets. This proved to be fifty feet long. In contrast, in other cities, cars of forty, forty-four, or forty-six feet were more common.

The same need caused the city to require acquisition and operation of trailers, also fifty feet long. A motor-trailer train, one hundred feet long, carried 250 passengers with a three-man crew. That was efficiency long before one-man cars were in use elsewhere. Trailers were very lightweight because they lacked propulsion equipment and were about half the weight of a motor car. With train operation, revenue per train mile could be adequate even with a three-cent fare.

The difference between the prevailing Cleveland Electric Railway five-cent fare and the three-cent fare was significant. Consider that a typical worker earned one dollar a day prior to World War I. A five-cent fare was a large chunk of his daily wage. Three cents was more tolerable. With the difference, he could buy a newspaper to read on the streetcar on his way home.

Very early, Cleveland had adopted single-end cars. They had more usable interior floor space, were simpler electrically, and cost less to buy than double-end cars; however, they did need a wye or loop terminal.

In 1910, the last wood cars (901–925) were delivered by Kuhlman to Cleveland Railway. They were fifty-two feet long and were converted to Peter Witt type cars in 1915 but never pulled trailers.

Under the direction of general manager, and later president, John J. Stanley, master mechanic Terrance Scullin prepared the design for a new generation of streetcars. Prototype 824 was built in 1912, a front- and rear-door fifty-foot car following the established Pay-as-You-Enter (PAYE) floor plan (with passengers entering at the rear door, exiting at the front door). Its roof featured a small clerestory on a high arch roof. The clerestory covered several vents in the roof and protected them from rainfall. It also produced airflow, like a wind tunnel, which created a vacuum above the vent and thus produced airflow in the car. The roof was patented by Mr. Scullin and was known thereafter as the Scullin Roof. It was used on car fleets 1000–1052, 1100–1300, 200–329, and 2300–2375, all built between 1911 and 1918.

The "ten-hundreds," built in 1913 to PAYE floorplan, had large (32- to 34-inch) wheels in Brill 27F trucks having outside hung motors. This permitted a short wheelbase, and thus the car was able to take curves with little screeching. The popular K-type manual controller was used, with straight air brakes. Rear couplers with automatic pneumatic tappet for the brake pipe were included as a standard part of Cleveland Railway motor cars to pull trailers. An outboard Westinghouse air compressor provided compressed air for the motor car and its trailer. A socket on the rear letterboard was provided for the trailer's cable that included a 600 volts DC cable for lighting, stove blower, and a separate wire for the conductor's signal buzzer. Car 1048 was specially built to use Brill 51E1 lowfloor trucks and retained them until retirement in 1946.

### Fleet Statistics: High-Floor "Ten-Hundreds," 1000–1052

BLDR: G. C. Kuhlman; LEN: 50' 2"; MOT: 4WH307F; doors: hand, fold; W: 47,900 lbs; signs: Hunter; seats: 55; TRKS: Brill 27 F-E-1; CTL: K35G; HDL: Dayton 1583A; CR T: 16 (1048, T 17; 1052, T 15). A parade was held 10-1-46 to say good-bye to the ten-hundreds and inaugurate PCC car service on St. Clair. The first car in line-up was work car 0140, with "In 1890 I was hot stuff" written on its side, followed by car 1020 with "Goodbye, Folks, I'm through," and then PCC Transitliners 4200 and 4201.

| | |
|---|---|
| 1000 | D 7-17-13, REN 0649 in 1946 for work car service, operated in "Parade of Progress" 4-27-52, SCR 2-26-53 |
| 1001 | D 7-9-13, forced air scoop on rear louver, SCR 9-11-47 |
| 1002 | D 7-9-13, SCR 9-15-47 |
| 1003 | D 7-10-13, placed in storage in 1940, SCR 4-24-41 |
| 1004 | D 7-10-13, placed in storage at Superior in 1940, SCR 2-9-41 |
| 1005 | D 7-11-13, REN 0648 in 1946 for work car service, SCR 3-22-54 |
| 1006 | D 7-11-13, placed in storage in 1940, SCR 2-12-41 |
| 1007 | D 7-12-13, SCR 9-15-47 |
| 1008 | D 7-12-13, OOS 1947 at E. 55th Station, SCR spring 1948 |
| 1009 | D 7-12-13, placed in storage in 1940, SCR 3-14-41 |
| 1010 | D 7-12-13, REN 0647 in 1946 for work car service, SCR 3-29-54 |

Car 1022, high-floor Scullin-roof car, G. C. Kuhlman 1913, CR type 16. *(Cleveland Railway photo, Blaine Hays Collection)*

1011    D 7-17-13, SCR 9-15-47

1012    D 7-17-13, placed in storage in 1940, SCR 2-22-41

1013    D 7-24-13, placed in storage in 1940, SCR 3-5-41

1014    D 7-24-13, placed in storage at Harvard in 1940, CON for storage of Way Department material at Brooklyn Station in 1941 and retrieved and renovated for passenger service during wartime in 1942, SCR 5-29-47

1015    D 7-23-13, SCR spring 1947

1016    D 7-23-13, OOS at E. 55th Station 1947, SCR spring 1948

1017    D 7-25-13, SCR 5-27-47

1018    D 7-25-13, SCR spring 1947

1019    D 7-25-13, placed in storage at St. Clair Car House 1940, SCR 2-14-41

1020    D 8-1-13, heavily REB and improved, SCR 5-16-47

1021    D 7-30-13, placed in storage 1940, SCR 4-17-41

1022    D 7-29-13, placed in storage 1940, SCR 3-13-41

1023    D 8-2-13, placed in storage 1940, SCR 4-17-41

1024    D 8-2-13, placed in storage 1940, SCR 4-15-41

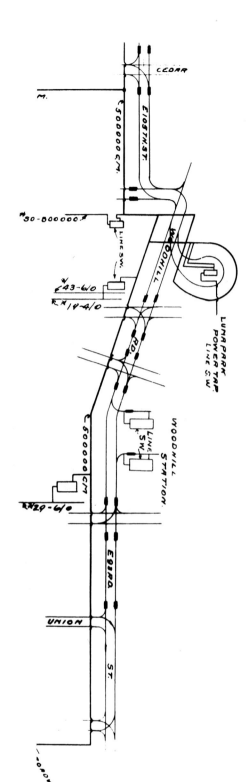

1025   D 8-4-13, placed in storage at Superior 1940, SCR 3-11-41

1026   D 8-1-13, placed in storage 1940, SCR 4-1-41

1027   D 8-4-13, placed in storage 1940, SCR 4-3-41

1028   D 8-1-13, placed in storage 1940, SCR 3-26-41

1029   D 8-6-13, placed in storage 1940, SCR 3-26-41

1030   D 8-6-13, placed in storage 1940, SCR 4-1-41 (also reported SCR 5-13-41)

1031   D 8-6-13, placed in storage 1940, SCR 3-4-41 (also reported SCR 3-7-41)

1032   D 8-7-13, SCR 5-20-47

1033   D 1913, OOS at Denison Station, SCR spring 1948

1034   D 8-8-13, placed in storage at Harvard 1940, SCR 2-5-41 (also reported SCR 3-1-41)

1035   D 8-12-13, placed in storage 1940, SCR 2-6-41 (also reported SCR 3-1-41)

1036   D 8-12-13, SCR 5-29-47

1037   D 8-14-13, placed in storage 1940, SCR 2-8-41 (also reported SCR 3-5-41)

1038   D 8-14-13, SCR 9-11-47

1039   D 8-19-13, OOS 1947, SCR 1948

1040   D 8-19-13, placed in storage 1940, SCR 4-4-41 (also reported SCR 5-15-41)

1041   D 8-20-13, placed in storage 1940, SCR 2-27-41 (also reported SCR 3-6-41)

1042   D 8-20-13, CON to storage of Way Department material at Madison Station in 1941 and retrieved and renovated for passenger service during wartime in 1942, SCR spring 1948

1043   D 8-21-13, OOS at Denison Station 1947, SCR spring 1948

1044   D 1913, car damaged in collision on Heights line 3-15-21, SCR 6-23-47

1045   D 8-28-13, CON for storage of Way Department material at St. Clair Station in 1941, REN 0650 1946 for work car service, SCR 5-17-54

1046   D 8-22-13, CON for storage of Way Department material at Woodhill Station 1941, SCR 7-23-47

1047   D 8-23-13, placed in storage 1940, SCR 4-4-41 (also reported SCR 5-14-41)

1048   D 8-26-13, CON to storage of Way Department materials at East 55th Station 1941, B51E-1 TRKS, SCR 5-9-47

1049   D 8-27-13, placed in storage at Superior 1940, SCR 3-10-41 (also reported SCR 3-12-41)

1050   D 9-3-13, in storage 1940, SCR 3-8-41 (also reported SCR 3-11-41)

1051   D 9-3-13, in storage at Harvard 1940, CON to storage of Way Department materials at Denison Station 1941, retrieved and renovated for passenger service during wartime 1942, CON into gardener's shanty in 1945 at Rocky River Station for care of former station superintendent William Allton's famous garden, OOS at Denison Station 1947, SCR spring 1948

1052  BLT CR 1911 as car 824, construction of car supervised by CR master mechanic Terrance Scullin, who designed and patented the louver roof, of which this was first application, car LEN 50' 5/8", in storage 1940, SCR 3-12-41 (also reported as SCR 3-14-41)

### Evolution of the Drop-Center Motor and Trailer
#### by J. William Vigrass

An experimental car, number 101, was received from Kuhlman by Cleveland Railway in 1912. With a drop-center platform where the conductor was located, it was modeled after the Hedley-Doyle stepless "hobble skirt" cars of New York Railways. To allow the low floor, Brill single-motor maximum traction trucks were used, with the pony wheels inward toward the center doors. The large 32-inch driving

Car 1107, as built, center-entrance type, G. C. Kuhlman, 1914, CR type 21. *(Cleveland Railway photo, Blaine Hays Collection)*

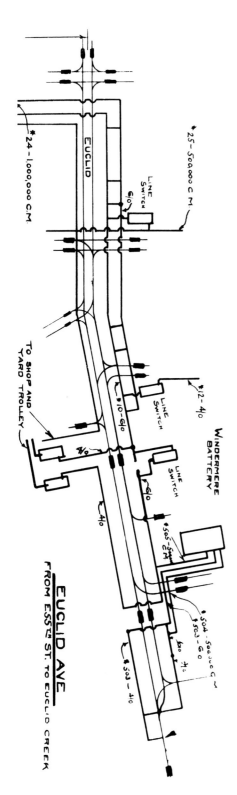

wheels faced outward. The car was not duplicated and spent its life as the Linndale Dinkey. However, the drop-center low-floor feature inspired the 1100-type cars and matching trailers that followed.

In 1913 and 1914, fifty new center-entrance, center-exit motor cars, type 1100–1149, were delivered by the G. C. Kuhlman Car Company, a subsidiary of the J. G. Brill Company, Cleveland Railway's regular supplier. The 1100s featured the new low-floor Westinghouse 340 motors, the same as car 1048, which demonstrated the low-floor truck. These cars had been designed to duplicate performance of the large motors used on previous cars having 34-inch wheels and high floors. They were initially rated at 40 hp, but later versions were rated at 50 hp. In any case, they were designed to pull trailers. They allowed 26-inch-diameter wheels that, in turn, allowed a low-floor car similar to the unpowered trailers. Low-floor motors appeared concurrently in other cities but were generally of a lower power. Cleveland's use of the WH340 motor was unique, and it had a long and successful relationship with Cleveland.

The 1100s were carried on Brill 51E-1 trucks, a short-wheelbase (4'10") truck with outside hung motors. It had a one-piece forged steel sideframe, a feature of Brill trucks. The interior floor plan had transverse cushioned rattan seats on the right side and a longitudinal hard rattan-covered bench on the left side.

The conductor's position was between the two center doors. The passenger took one step from the street to the drop-center platform of the car and then stepped up two steps to the main seating area of the car. A few seats (three or four) were opposite the conductor's stand in the drop center, as was the coal stove. The latter could be removed in summer to provide another seat space (though in later years, it was not removed).

The motorman had an enclosed cab with an electric heater all to himself where he could survey the street ahead in solitary splendor, which was especially appreciated in winter.

The Cleveland Railway provided a place to work for the motorman, but it did not provide a place to sit down! If a motorman wished to sit, he had to buy and maintain his own stool (except on PCC cars, which had a built-in seat). To store stools, the company provided in each carbarn a pipe rack with large pigeon holes, and stools were chained to the rack.

The fifty new 1100s were soon followed by fifty more, 1150–1199, also in 1914. They had longitudinal bench seats throughout and matched trailers in their basic floorplan, but they proved unpopular with passengers and were not repeated. In later years the 1150s saw use mostly in rush hours.

In 1914 and 1915, five more orders were placed for 101 more center-door cars, 1200, 1201–1225, 1226–1250, 1251–1275, and 1276–1300. They were nearly identical to 1100–1149 but had a few improvements

Car 1289, as rebuilt, center-entrance type, G. C. Kuhlman, 1915, CR type 21. *(Cleveland Transit photo, Blaine Hays Collection)*

based on experience, one being a somewhat more reliable pneumatic door actuator, the other being cross seats on one side of the aisle.

### Fleet Statistics: Center-Entrance Cars, 1100–1300

BLDR: G. C. Kuhlman Car Co.; roof: Scullin arch; MOT: 4WH340 (rapid transit type 340A-1); LEN: 51'; doors: air slide; signs: Hunter; W: 44,280 lbs.; seats: 58 (longitudinal cars, 54); TRKS: Brill 51-E-1 (rapid transit type B68E-1); CTL: K35G (rapid transit, multiple-unit HL); HDL: Dayton 1583A; city cars: CR T 21; rapid cars CR T 22 (disposition of rapid transit type continued in rapid transit roster).

| | |
|---|---|
| 1100 | D 12-27-13, wrecked on Superior 12-22-20, SCR 2-12-48 |
| 1101 | D 12-31-13, damaged in ACC on Superior 9-22-21, SCR 2-12-48 |
| 1102 | D 12-31-13, SCR 2-26-48 |
| 1103 | D 12-31-13, SCR 9-18-47 |
| 1104 | D 12-31-13, SCR 11-19-47 |

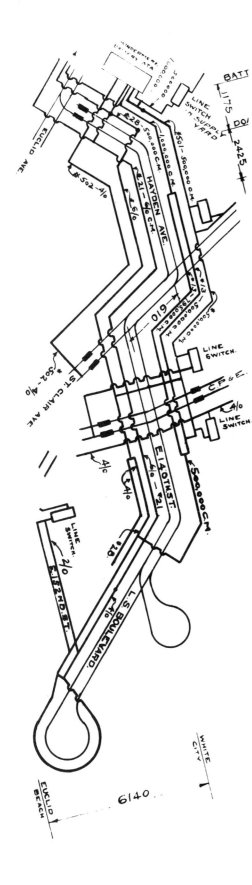

1105 D 4-5-14, ACC on Superior 9-22-21, SCR 11-29-47
1106 D 1-15-14, SCR 2-12-48
1107 D 1-15-14, SCR 2-12-48
1108 D 1-15-14, used for salt storage, SCR 3-30-48
1109 D 1-15-14, used for salt storage, SCR 9-15-47
1110 D 4-5-14, used for salt storage, SCR 9-4-47
1111 D 2-21-14, SCR 12-30-47
1112 D 3-21-14, used for salt storage, SCR 3-23-48
1113 D 1-22-14, SCR 2-26-48
1114 D 1-22-14, SCR 10-31-47
1115 D 1-22-14, SCR 12-10-47
1116 D 1-22-14, SCR 1-29-47
1117 D 1-1914, SCR 12-30-47
1118 D 2-13-14, SCR 12-17-47
1119 D 3-22-14, SCR 12-9-47
1120 D 4-1-14, SCR 2-2-48
1121 D 1914, SCR 2-25-48
1122 D 1914, used for salt storage, had painted on side toward end of service "no more jars on these cars," SCR 3-5-48
1123 D 1914, SCR 11-24-47
1124 D 3-20-14, SCR 1-21-48 (inventory says 1-21-47)
1125 D 3-1-14, SCR 12 17-47
1126 D 3-20-14, SCR 1-21-48 (inventory says 1-21-47)
1127 D 3-16-14, used for salt storage, SCR 1-15-48
1128 D 3-10-14, used for salt storage, SCR 3-30-48
1129 D 3-10-14, SCR 12-5-47
1130 D 3-23-14, used for salt storage, SCR 3-5-48
1131 D 3-6-14, SCR 12-5-47
1132 D 3-14-14, had painted on side "CTS is handing you a new line" in 1947, used for salt storage, SCR 1-9-48
1133 D 3-6-14, SCR 1-6-48
1134 D 1914, car rammed rear of Southwestern interurban at E. 21st and Superior 3-22-22, SCR 11-12-47
1135 D 1914, SCR 12-12-47
1136 D 1914, SCR early 1947
1137 D 1914, SCR 4-30-47
1138 D 3-19-14, SCR 9-25-47
1139 D 2-21-14, SCR 11-24-47
1140 D 2-21-14, ACC damage 1937, SCR 3-30-40
1141 D 3-7-14, SCR 12-10-47
1142 D 2-25-14, used for salt storage, SCR 12-24-47
1143 D 4-1-14, SCR 3-28-40
1144 D 4-1-14, SCR 3-26-40
1145 D 3-31-14, SCR 2-26-48
1146 D 3-27-14, SCR 12-2-47
1147 D 3-25-14, SCR 12-2-48

| | |
|---|---|
| 1148 | D 3-25-14, used for salt storage, SCR 3-23-48 |
| 1149 | D 3-30-14, SCR 12-2-47 |
| 1150 | D 4-21-14, used for salt storage, SCR 3-30-48 |
| 1151 | D 4-21-14, SCR 11-19-47 |
| 1152 | D 4-21-14, SCR 9-18-47 |
| 1153 | D 4-30-14, collision at Lorain Station 7-16-22, used for salt storage, SCR 1-16-48 |
| 1154 | D 5-1-14, used for salt storage, SCR 3-5-48 |
| 1155 | D 5-2-14, SCR 1-29-48 |
| 1156 | D 5-2-14, used for salt storage, SCR 3-5-48 |
| 1157 | D 5-2-14, SCR 12-5-47 |
| 1158 | D 5-5-14, SCR 11-19-47 |
| 1159 | D 5-5-14, damaged in collision under high-level bridge 8-13-21, ACC Cedar Glen 1 dead and 14 injured on 9-4-43, SCR 11-24-47 |
| 1160 | D 5-5-14, SCR 1-21-48 (inventory says 1-21-47) |
| 1161 | D 1914, damaged in collision on Public Square 12-19-22, SCR 3-23-40 |
| 1162 | D 1914, used for salt storage, SCR 3-30-48 |
| 1163 | D 1914, SCR 11-21-47 |
| 1164 | D 1914, had GE275 MOT, SCR 2-12-48 |
| 1165 | D 1914, damaged by fire, OSS since 1930, stored at Miles Station for ten years in original paint, SCR 3-29-40 |
| 1166 | D 1914, SCR 7-21-47 |
| 1167 | D 1914, SCR 12-12-47 |
| 1168 | D 1914, damaged in collision at Fulton and Lorain 7-16-22, SCR 2-25-48 |
| 1169 | D 5-20-14, SCR 2-2-48 |
| 1170 | D 5-20-14, SCR 11-12-47 |
| 1171 | D 5-20-14, SCR 11-12-47 |
| 1172 | D 5-20-14, used for salt storage, SCR 12-24-47 |
| 1173 | D 5-20-14, SCR 11-29-47 |
| 1174 | D 5-20-14, used for salt storage, SCR 3-23-48 |
| 1175 | D 5-20-14, damaged in collision at Fulton and Lorain 7-16-22, SCR 2-2-48 |
| 1176 | D 5-20-14, SCR 12-2-47 |
| 1177 | D 5-28-14, SCR 3-22-40 |
| 1178 | D 5-28-14, ACC on Prospect 12-23-20, SCR 10-2-47 |
| 1179 | D 5-28-14, damaged in collision at Broadway and Harvard 4-5-22, used for salt storage, SCR 1-15-47 |
| 1180 | D 5-28-14, used for salt storage, SCR 8-19-47 |
| 1181 | D 5-28-14, SCR 4-25-47 |
| 1182 | D 5-28-14, used for salt storage, SCR 3-23-48 |
| 1183 | D 6-5-14, SCR 9-23-47 |
| 1184 | D 6-5-14, used for salt storage, SCR 3-30-48 |
| 1185 | D 6-6-14, SCR 11-26-47 |

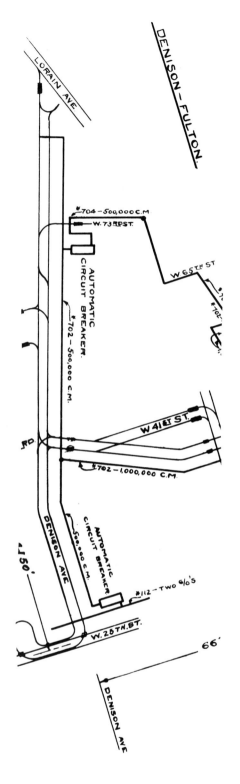

| | |
|---|---|
| 1186 | D 6-9-14, used for salt storage, SCR 7-2-47 |
| 1187 | D 6-9-14, collision on Broadway 12-18-22, SCR 3-23-46 |
| 1188 | D 6-9-14, SCR 11-24-47 |
| 1189 | D 6-15,14, SCR 12-5-47 (inventory says 11-24-47) |
| 1190 | D 6-15-14, SCR 7-16-47 |
| 1191 | D 6-15-14, SCR 12-2-47 |
| 1192 | D 6-20-14, SCR 1-29-48 |
| 1193 | D 6-20-14, SCR 9-18-47 |
| 1194 | D 6-20-14, collision at Champlain and Ontario 7-16-22, burned to the frame and stored 1934, SCR 3-8-40 |
| 1195 | D 7-1-14, SCR 9-25-47 |
| 1196 | D 7-1-14, SCR 5-5-47 |
| 1197 | D 7-16-14, SCR 12-10-47 |
| 1198 | D 7-17-14, ACC on Union 12-22-20, SCR 5-7-47 |
| 1199 | D 7-18-14, collision at Prospect and Ontario 7-26-22, SCR 3-21-40 |
| 1200 | D 3-2-15, FN 1300 (REN 1922), sent to Shaker Rapid in 1920, sold to CIRR 4-15-40, moved to CIRR 5-10-40 |
| 1201 | D 3-1-15, FN 1299 (REN 1922), sent to Shaker Rapid in 1920, sold to CIRR 4-15-40 |
| 1202 | D 2-26-15, FN 1298 (REN 1922), sent to Shaker Rapid in 1920, sold to CIRR 4-15-40 |
| 1203 | D 10-17-14, TRKS Standard C50, sent to Shaker Rapid in 1920, sold to CIRR 4-15-40 moved to CIRR 5-21-40 |
| 1204 | D 10-20-14, sent to Shaker Rapid in 1920, sold to CIRR 4-15-40, moved to CIRR 5-17-40 |
| 1205 | D 10-20-14, sent to Shaker Rapid in 1920, sold to CIRR 4-15-40 |
| 1206 | D 10-20-14, sent to Shaker Rapid in 1921, sold to CIRR 4-15-40, moved to CIRR 5-14-40 |
| 1207 | D 10-28-14, sent to Shaker Rapid in 1921, sold to CIRR 4-15-40, moved to CIRR 5-21-40 |
| 1208 | D 10-31-14, TRKS Standard C50, sent to Shaker Rapid in 1921, sold to CIRR 4-15-40 |
| 1209 | D 10-29-14, sent to Shaker Rapid in 1921, sold to CIRR 4-15-40 |
| 1210 | D 10-29-14, TRKS Standard C50, sent to Shaker Rapid in 1921, sold to CIRR 4-15-40 |
| 1211 | D 10-29-14, sent to Shaker Rapid 1921, sold to CIRR 4-15-40 |
| 1212 | D 10-29-14, sent to Shaker Rapid 1921, sold to CIRR 4-15-40 |
| 1213 | D 10-31-14, sent to Shaker Rapid 1921, sold to CIRR 4-15-40 |
| 1214 | D 10-31-14, TRKS Standard C50, sent to Shaker Rapid 1921, sold to CIRR 4-15-40 |
| 1215 | D 11-5-14, sent to Shaker Rapid 1921, sold to CIRR 4-15-40 |
| 1216 | D 11-9-14, sent to Shaker Rapid 1921, sold to CIRR 4-15-40 |
| 1217 | D 11-5-14, sent to Shaker Rapid 1921, sold to CIRR 4-15-40 |

1218    D 11-6-14, sent to Shaker Rapid 1921, sold to CIRR 4-15-40
1219    D 11-6-14, sent to Shaker Rapid 1921, sold to CIRR 4-15-40
1220    D 11-7-14, sent to Shaker Rapid 1925, sold to CIRR 4-15-40
1221    D 11-7-14, sent to Shaker Rapid 1925, sold to CIRR 4-15-40
1222    D 11-10-14, sent to Shaker Rapid 1925, sold to CIRR 4-15-40
1223    D 11-10-14, sent to Shaker Rapid 1925, sold to CIRR 4-15-40
1224    D 11-10-14, sent to Shaker Rapid 1925, sold to CIRR 4-15-40
1225    D 11-10-14, sent to Shaker Rapid 1925, sold to CIRR 4-15-40, moved to CIRR 5-17-40
1226    D 11-18-14, sent to Shaker Rapid 1925, sold to CIRR 4-15-40
1227    D 11-18-14, sent to Shaker Rapid 1925, sold to CIRR 4-15-40
1228    D 11-18-14, sent to Shaker Rapid 1925, sold to CIRR 4-15-40
1229    D 11-21-14, sent to Shaker Rapid 1925, sold to CIRR 4-15-40
1230    D 11-21-14, sent to Shaker Rapid 1925, sold to CIRR 4-15-40
1231    D 11-21-14, sent to Shaker Rapid 1925, sold to CIRR 4-15-40
1232    D 11-21-14, collision at Prospect and Ontario 7-26-22, sent to Shaker Rapid 1925, sold to CIRR 4-15-40
1233    D 11-23-14, CON to trailer in 1927, ACQ 51-E-1 TRKS, sent to Shaker Rapid in 1927, sold to CIRR 4-15-40
1234    D 11-23-14, collision on Broadway 12-18-22, CON to trailer in 1927, ACQ B51-E-1 TRKS, sent to Shaker Rapid in 1927, sold to CIRR 4-15-40, moved to CIRR 5-22-40
1235    D 11-24-14, CON to trailer in 1927, ACQ B51-E-1 TRKS, sent to Shaker Rapid in 1927, sold to CIRR 4-15-40, moved to CIRR 5-28-40
1236    D 11-27-14, used for salt storage, SCR 4-27-48
1237    D 11-27-14, used for salt storage, SCR 5-20-48
1238    D 11-27-14, used for salt storage, SCR 5-20-48
1239    D 11-27-14, used for salt storage, SCR 4-27-48
1240    D 11-28-14, used for salt storage, SCR 4-27-48
1241    D 12-2-14, used for salt storage, SCR 4-27-48
1242    D 12-2-14, used for salt storage, SCR 4-27-48
1243    D 12-2-14, SCR 4-27-48
1244    D 12-2-14, used for salt storage, SCR 4-27-48
1245    D 12-2-14, used for salt storage, SCR 4-27-48
1246    D 12-11-14, turned on side in ACC at bottom of Cedar Hill 12-10-37, used for salt storage, SCR 5-3-48
1247    D 12-12-14, used for salt storage, SCR 5-3-48
1248    D 12-12-14, motors removed 10-5-37, ACC on Clifton 1938, SCR 3-25-40
1249    D 12-12-14, used for salt storage, SCR 4-27-48
1250    D 12-12-14, used for salt storage, SCR 5-3-48
1251    D 12-18-14, SCR 1-29-48
1252    D 12-18-14, SCR 11-26-47
1253    D 12-19-14, CON to farebox storage car at East 55th Station, SCR 4-30-48

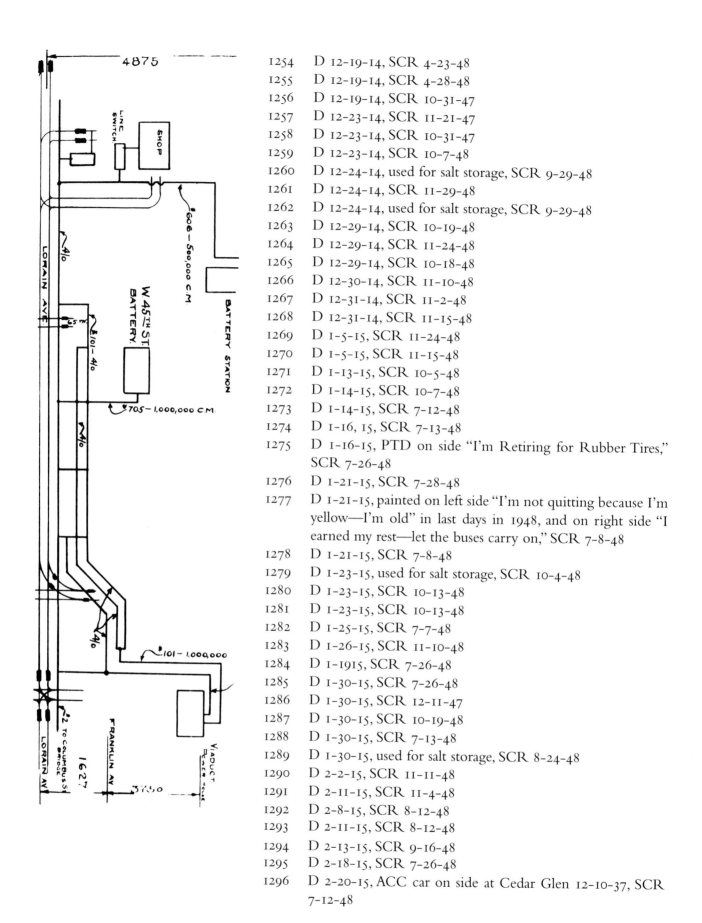

| | |
|---|---|
| 1254 | D 12-19-14, SCR 4-23-48 |
| 1255 | D 12-19-14, SCR 4-28-48 |
| 1256 | D 12-19-14, SCR 10-31-47 |
| 1257 | D 12-23-14, SCR 11-21-47 |
| 1258 | D 12-23-14, SCR 10-31-47 |
| 1259 | D 12-23-14, SCR 10-7-48 |
| 1260 | D 12-24-14, used for salt storage, SCR 9-29-48 |
| 1261 | D 12-24-14, SCR 11-29-48 |
| 1262 | D 12-24-14, used for salt storage, SCR 9-29-48 |
| 1263 | D 12-29-14, SCR 10-19-48 |
| 1264 | D 12-29-14, SCR 11-24-48 |
| 1265 | D 12-29-14, SCR 10-18-48 |
| 1266 | D 12-30-14, SCR 11-10-48 |
| 1267 | D 12-31-14, SCR 11-2-48 |
| 1268 | D 12-31-14, SCR 11-15-48 |
| 1269 | D 1-5-15, SCR 11-24-48 |
| 1270 | D 1-5-15, SCR 11-15-48 |
| 1271 | D 1-13-15, SCR 10-5-48 |
| 1272 | D 1-14-15, SCR 10-7-48 |
| 1273 | D 1-14-15, SCR 7-12-48 |
| 1274 | D 1-16, 15, SCR 7-13-48 |
| 1275 | D 1-16-15, PTD on side "I'm Retiring for Rubber Tires," SCR 7-26-48 |
| 1276 | D 1-21-15, SCR 7-28-48 |
| 1277 | D 1-21-15, painted on left side "I'm not quitting because I'm yellow—I'm old" in last days in 1948, and on right side "I earned my rest—let the buses carry on," SCR 7-8-48 |
| 1278 | D 1-21-15, SCR 7-8-48 |
| 1279 | D 1-23-15, used for salt storage, SCR 10-4-48 |
| 1280 | D 1-23-15, SCR 10-13-48 |
| 1281 | D 1-23-15, SCR 10-13-48 |
| 1282 | D 1-25-15, SCR 7-7-48 |
| 1283 | D 1-26-15, SCR 11-10-48 |
| 1284 | D 1-1915, SCR 7-26-48 |
| 1285 | D 1-30-15, SCR 7-26-48 |
| 1286 | D 1-30-15, SCR 12-11-47 |
| 1287 | D 1-30-15, SCR 10-19-48 |
| 1288 | D 1-30-15, SCR 7-13-48 |
| 1289 | D 1-30-15, used for salt storage, SCR 8-24-48 |
| 1290 | D 2-2-15, SCR 11-11-48 |
| 1291 | D 2-11-15, SCR 11-4-48 |
| 1292 | D 2-8-15, SCR 8-12-48 |
| 1293 | D 2-11-15, SCR 8-12-48 |
| 1294 | D 2-13-15, SCR 9-16-48 |
| 1295 | D 2-18-15, SCR 7-26-48 |
| 1296 | D 2-20-15, ACC car on side at Cedar Glen 12-10-37, SCR 7-12-48 |

74 · CLEVELAND'S TRANSIT VEHICLES

1297  D 2-11-15, SCR 7-7-48

1298  D 10-17-14, FN 1202 (REN 1922), "farewell to rails—buses are coming Saturday," PTD on side in 1948, SCR 7-7-48

1299  D 10-17-14, used for salt storage, SCR 10-4-48

1300  D 8-8-14, car cost $10,417.00, SCR 12-5-47

### The Peter Witt Car
*by J. William Vigrass*

Narrow-door-loading motor and trailer cars suffered from slow boarding situations in peak times. In an effort to reduce dwell time at stops and speed up loading, Commissioner Peter Witt, ever alert to system needs, conceived and patented the "Car Rider's Car," which came to bear his name. It featured a double front door and double center doors and created the "pay-as-you-pass" fare collection system. The entire front half of the car was an unpaid area. It had longitudinal rattan covered benches on both sides. A passenger could then pay his or her fare and exit, or pay the fare and proceed to the rear half of the car, where 2-2 transverse cushioned rattan seats were provided to entice passengers to move to the rear.

The prototype "Car Rider's Car," which came to be known to the industry as the Peter Witt, was car 33 (later renumbered 330), built by Cleveland Railway in 1914. It was followed later that year by Kuhlman-built production cars 200–249 and in 1916 by 250–329. They featured a 44-inch front door reported in the trade press by the builder to be suitable for a double stream of boarding passengers. Two adjacent center doors, each 33 inches wide, were aft of the conductor's position. The 50-foot car seated fifty-five. Witt never charged Cleveland Railway any money for the use of his design, in spite of the hundreds of cars built for the city.

Each car built under the patented Peter Witt floor plan carried a small brass plate (1" x 3") above one of the center doors upon which was engraved: "Car Rider's Car Patented April 25, 1916."

The width of the front doors became an issue with construction of the 200s. The 44-inch front door they were built with was not entirely satisfactory. Two cars, 225 and 291, had a wider door installed by removing the adjacent posts, making it about a foot wider. No other 200s were so modified, so apparently it was not worth the cost. Later, Peter Witt cars of the 350, 400, and 4000 series had a very wide front door of 56 inches. Yet, the 200s soldiered on until retirement with a 44-inch door.

Approximately enough of the 200 series "crosstown" cars, 200–207 were converted to one-man in 1924–26 for the lightly traveled East 30th Street crosstown line. They had treadle-operated center doors, an open cab for the one-man operator (with no partition with door as

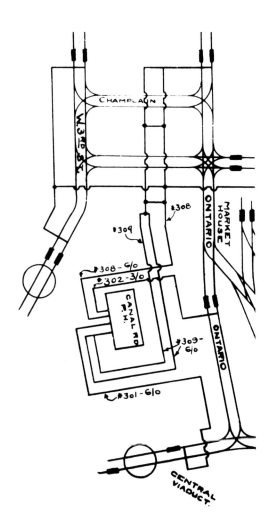

the motorman of a two-man car had). Yet they retained their coal stoves. They operated as one-man cars until November 1941, when East 30th Street was converted to motorbus. They were reconverted to two-man configuration during World War II and operated in that mode until retired.

Following the successful introduction of the 200s, Witt recommended that some existing cars be modified to the new plan. Accordingly, the railway selected fifty wood cars from its fleet and planned to modernize them. Only twenty-six cars were in fact modernized: 900 and 901–925, and a second 920 was rebuilt from open car 415 in 1918. No other open cars were done, even though twenty-five had been planned.

The 900s were built as a 52-foot pay-as-you-enter (PAYE) car. After conversion, they were 49 feet 3⅜ inches long. Former open car 415 had concave sides and rub rails. The other 900s had flat sides and rub rails and were typical of cars built from 1890 to 1914. The concave sides cleared wagon wheel hubs, and rub rails protected carefully painted car sides from abrasion by wagons. Traffic congestion by this time was becoming severe, and nearly all economic activity was located along streetcar lines, employment concentrated along car lines by necessity. The rebuilt 900s were unique as being the only high-wheeled Peter Witts ever built. No further rebuilding of old-type cars to Peter Witts was undertaken, as new cars were preferred.

Following the successful introduction of 200–249, and rebuilt 900–925, the railway began a systematic accumulation of a large fleet of Peter Witt motor cars. While body styles changed somewhat with each order, the technology used was remarkably uniform. Standardization and economies of scale carried efficiency to a high level.

The second order of 200s soon followed, 250–329, and they were identical to 200–249 with their 44-inch front door. After World War I, motor car orders resumed in a big way in 1920 with 450–499, a design that can be considered the classic Peter Witt. They were originally assigned with their matching trailers 2451–2500 to Euclid Avenue, Cleveland's main street and busiest line. The conductor's stand was between the center doors, unlike 200s and older trailers that had adjacent center doors with the conductor's position forward of them. This fleet was ordered with Brill 68-E trucks having inside-hung motors and a longer wheelbase of six feet. The WH340 motor and related K-type manually operated controller on the platform were employed. Ohio Brass Company's Tomlinson-type coupler, flat face with hook, with integral pneumatic tappet, completed the car, equipping it to pull trailers. Doors were pneumatically operated and matching trailers 2451–2500 provided trains. The 450s featured an outward-folding wider front door of 56 inches, two generous patron streams wide. The 56-inch width solved the problem of boarding large crowds rapidly. Subsequent cars retained the width but had inward-folding front doors

Slotted floors, rattan seats, and wooden handrails with brass fittings were standard features of early day Cleveland streetcars. *(Cleveland Railway photo, Blaine Hays Collection)*

and eliminated the folding step in favor of fixed steps. Sixteen (484–499) of the 450s lost their six-foot 68-E trucks to Shaker Rapid 1200s and, in turn, received short wheelbase 4'10" 51-Es.

Cars 350–449 with matching trailers (2376–2425) soon followed in 1922 and 1923. They had mechanically operated doors operated by manual hand cranks. This appears to have been to enhance reliability because pneumatic door cylinders sometimes leaked (gaskets failed) or would freeze in winter.

In 1924 and 1925, the Cleveland Railway built cars 150–174 at Harvard Shops to the floor plan of the 200s, with adjacent center doors but with a wide 56-inch front door. Mechanical door operators as in the 350s were used. They had a plain arch roof with ventilators, a more modern appearing roof than the high arch Scullin roof. Nonetheless, the 150s reflected Scullin's styling.

The 150s were followed in 1925 by 1301–1375, also built by the Railway, which were similar but had the conductor's station between center doors. Mechanical door operators provided simple, reliable operation. Thus, the 150s plus 1301s provided one hundred more Peter Witt Car Rider's Cars. Sixteen matching trailers (1982–1997) were also built in the shops in 1925.

Car 1376, built in 1926 by Cleveland Railway, was notable in that it was built entirely of aluminum, including the truck castings. It was featured at the 1926 American Electric Railway Association convention at Cleveland's Public Auditorium. Full-page ads by ALCOA appeared in the *Electric Railway Journal*. This lightweight car was equipped with a 35 hp motor and was found to be unsuitable for pulling trailers, even though originally equipped with a coupler. Aside from its unpainted aluminum handrails and interior fittings, it looked like any other 1301. It operated until 1951, having had a full service life.

The fleets of 4000s and 5000s, built between 1927 and 1930, were the total refinement of the Peter Witt concept. They were the busiest and most well-remembered of the streetcar fleets.

### Fleet Statistics: Peter Witt, 330

World's first Peter Witt car—BLDR: CR, Lakeview Shops; LEN: 51' 1$^{5}/_{16}$"; W: 46,760 lbs.; seats: 56; TRKS: Brill 51 E-1; MOT: 4WH340; CTL: K35G; HDL: Dayton 1583A; FEN: Eclipse; roof: arch; CR T 23.

330   FN 33, D 12-1-14, SCR 10-24-46 (also listed as SCR 5-20-47)

### Fleet Statistics: Peter Witt, 200–299

BLDR: G. C. Kuhlman; LEN: 51' 1"; W: 45,560 lbs; seats: 55; 1-man, 10 cross (all converted back to 2-man in 1941); 2-man, 12 cross and longitudinal; TRKS: Brill 51-E-1; MOT: 4WH340; CTL: K35G; HDL: Dayton 1583A; FEN: Eclipse; roof: arch; CR T 23.

| | |
|---|---|
| 200 | D 7-29-15, CON OM 6-8-26, OOS E. 55 1948 and SCR 11-9-50 |
| 201 | D 8-1-15, CON OM 6-18-26, OOS 1948 and SCR 6-1-48 |
| 202 | D 8-1-15, CON OM 6-19-26, OOS 1948 and SCR 10-1-48 |
| 203 | D 8-6-15, CON OM 7-9-26, OOS E. 55 1948 and SCR 11-9-50 |
| 204 | D 8-6-15, CON OM 6-30-26, OOS 1948 and SCR 12-31-48 |
| 205 | D 8-7-15, CON OM 6-28-26, OOS 1948 and SCR 9-10-48 |
| 206 | D 8-8-15, CON OM 6-30-26, OOS E. 55th 1950 and SCR 1-23-52 |
| 207 | D 8-8-15, CON OM 7-3-26, OOS 1948 and SCR 6-1-48 |
| 208 | D 8-8-15, OOS 1949 for CON to Tow Car 0208, operated in "Parade of Progress" 4-27-52, SCR 10-7-53 |
| 209 | D 8-12-15, OOS E. 55th 1950 and SCR 1-3-52 |

Car 330, FN 33, was the first Peter Witt streetcar, Cleveland Railway 1914, CR type 23. *(Cleveland Railway photo, Blaine Hays Collection)*

Car 290 (in green paint), Peter Witt, G. C. Kuhlman, 1916, CR type 23. *(Cleveland Railway photo, Blaine Hays Collection)*

| | |
|---|---|
| 210 | D 8-13-15, OOS E. 55th 1948 and SCR 5-12-51 |
| 211 | G D 8-13-15, OOS 1948 and SCR 6-23-48 |
| 212 | D 8-18-15, OOS 1947 and SCR 12-11-47 |
| 213 | D 8-13-15, OOS 1948 and SCR 9-10-48 |
| 214 | D 8-13-15, OOS 1948 and SCR 5-28-48 |
| 215 | D 8-13-15, OOS E. 55th 1950 and SCR 1-23-52 |
| 216 | D 8-18-15, OOS 1947 and SCR 12-17-47 |
| 217 | D 8-18-15, OOS 1949 and SCR 2-17-49 |
| 218 | D 8-18-15, OOS 1949 and SCR 1-24-49 |
| 219 | D 8-20-15, OOS 1951 and SCR 11-9-51 |
| 220 | D 8-20-15, OOS 1947, used for salt storage, SCR 3-23-48 |
| 221 | D 8-21-15, OOS 1949 and SCR 3-1-49 |
| 222 | D 8-23-15, OOS 1948 and SCR 9-16-48 |
| 223 | G, D 8-23-15, OOS 1948 and SCR 6-23-48 |
| 224 | D 8-25-15, OOS 1948 and SCR 8-19-48 |
| 225 | D 8-31-15, had extra-wide doors for fast loading, OOS 1949 and SCR 3-1-49 |
| 226 | G, D 8-27-15, OOS 1946? and SCR 9-10-47 |
| 227 | D 9-3-15, OOS 1949 and SCR 1-24-49 |
| 228 | G, D 9-4-15, OOS 1949 and SCR 10-31-49 |
| 229 | G, D 9-4-15, OOS 1949 and SCR 1-3-49 |
| 230 | D 9-4-15, OOS 1948 and SCR 7-26-48 |
| 231 | D 9-4-15, OOS 1950 and SCR 10-10-50 |
| 232 | D 9-9-15, OOS 1949 and SCR 3-14-49 |
| 233 | D 9-9-15, OOS E. 55th 1950 and SCR 10-25-51 |
| 234 | D 9-11-15, OOS E. 55th 1948 and SCR 10-10-50 |
| 235 | G, D 9-11-15, OOS 1948 and SCR 6-9-48 |
| 236 | D 9-13-15, OOS E. 55th 1948 and SCR 10-21-49 |
| 237 | G, D 9-15-15, ACC Buckeye Road Hill on 12-1-43, OOS 1949 and SCR 2-18-49 |
| 238 | D 9-16-15, OOS E. 55th 1950 and SCR 8-27-51 |
| 239 | D 9-22-15, OOS E. 55th 1950 and SCR 9-10-51 |
| 240 | D 9-22-15, OOS 1951 and SCR 1-23-52 |
| 241 | D 9-22-15, OOS 1949 and SCR 5-12-49 |
| 242 | G, D 9-24-15, OOS 1950 and SCR 11-16-50 |
| 243 | D 9-25-15, OOS 1948 and SCR 9-20-48 |
| 244 | G, D 10-1-15, OOS 1948 and SCR 9-10-48 |
| 245 | G, D 10-2-15, OOS 1949 and SCR 2-16-49 |
| 246 | D 10-2-15, OOS E 55th 1948 and SCR 10-25-49 |
| 247 | D 10-12-15, OOS 1947 and SCR either 1-27-47 or 48? |
| 248 | D 10-13-15, OOS E. 55th 1950 and SCR 9-10-51 |
| 249 | G, D 10-28-15, OOS 1948 and SCR 9-20-48 |
| 250 | G, D 1-13-16, OOS E. 55th 1950 and SCR 12-5-51 |
| 251 | D 1-18-16, OOS E. 55th 1950 and SCR 9-12-51 |

252  D 1-18-16, OOS E. 55th 1948 and SCR 10-18-49
253  D 1-19-16, OOS E. 55th 1950 and SCR 12-5-51
254  D 1-19-16, OOS 1949 and SCR 11-8-49
255  D 1-21-16, OOS 1948 and SCR 7-7-48
256  G, D 1-22-16, OOS E. 55th 1950 and SCR 1-3-52
257  G, D 1-24-16, OOS 1948, used for salt storage, SCR 10-4-48
258  D 1-24-16, OOS E. 55th 1948 (file says SCR 11-2-49, inventory 9-12-51)
259  D 1-26-16, OOS E. 55th 1950 and SCR 8-21-51
260  D 1-26-16, 3-14-20, damaged on E. 55th collision, 9-21-21, splits switch on E. 49th, PTD red in 1935 and became portable medical office for the Eucharistic Congress at Cleveland Stadium, OOS 1951 and SCR 1951
261  G, D 1-28-16, OOS 1951 and SCR 10-15-51
262  G, D 1-28-16, OOS 1948, used for salt storage, SCR 8-24-48
263  G, D 1-29-16, OOS 1949 and SCR 2-10-49
264  G, D 1-31-16, OOS 1951 and SCR 1-17-52
265  G, D 2-1-16, OOS 1951 and SCR 3-18-52
266  D 2-1-16, OOS 1951 and SCR 10-23-51
267  D 2-1-16, OOS 1951 and SCR 11-21-51
268  D 2-3-16, OOS E. 55th 1948 and SCR 10-24-49
269  D 2-3-16, OOS 1949 and SCR 9-10-48
270  D 2-3-16, OOS E. 55th 1950 and SCR 8-17-51
271  G, D 2-4-16, OOS 1949 and SCR 1-14-49
272  D 2-4-16, OOS 1951 and SCR 10-25-51
273  D 2-5-16, OOS 1949 and SCR 1-13-49
274  D 2-7-16, OOS 1948 and SCR 6-9-48
275  G, D 2-8-16, OOS 1949 and SCR 1-6-49
276  D 2-8-16, OOS 1951 and SCR 2-15-52
277  G, D 2-10-16, OOS 1947 and SCR 12-17-47
278  D 2-10-16, OOS 1947 and SCR 1-21-48
279  G, D 2-11-16, OOS 1949 and SCR 1-14-49
280  D 2-11-16, split switch on E. 49th 9-21-21, OOS 1951 and SCR 1-29-51
281  G, D 2-11-16, OOS 1948 and SCR 9-20-48
282  G, D 2-12-16, OOS E. 55th 1948 and SCR 10-20-49
283  D 2-14-16, OOS 1951 and SCR 11-1-51
284  D 2-14-16, OOS 1951 and SCR 10-29-51
285  D 2-15-16, OOS 1951 and SCR 1951
286  D 2-26-16, OOS E. 55th 1948 and SCR 12-9-49
287  D 2-17-16, ice scrapers installed on front TRK, OOS 1950, SCR 4-10-51
288  D 2-18-16, OOS E. 55th 1948 and SCR 11-1-49
289  D 2-18-16, OOS 1951 and SCR 10-8-51

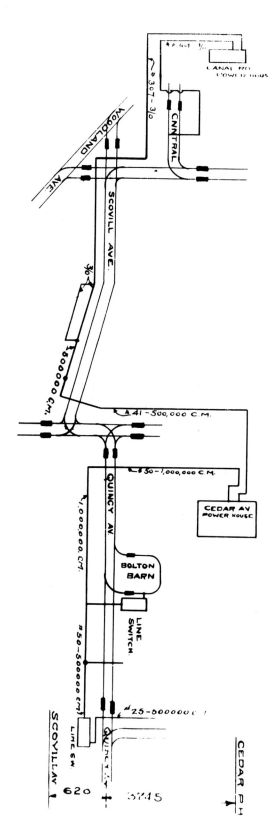

| | |
|---|---|
| 290 | G, D 2-19-16, OOS 1949 and SCR 2-8-49 |
| 291 | G, D 2-19-16, REB with larger front doors, OOS 1949, SCR 1-14-49 |
| 292 | D 2-29-16, OOS 1951 and SCR 10-23-51 |
| 293 | D 2-25-16, OOS E. 55th 1948 and SCR 10-25-49 |
| 294 | G, D 2-26-16, wrecked at E. 55th and Truscon 5-7-51, stored, SCR 11-12-51 |
| 295 | D 2-26-16, split switch 9-21-21 on E. 49, OOS 1950, SCR 7-17-50 |
| 296 | D 2-28-16, OOS E 55th 1949 and SCR 11-14-49 |
| 297 | D 2-29-16, OOS 1951 and SCR 1-1951 |
| 298 | G, D 3-1-16, damaged E. 55th collision 3-14-20, OOS 1949, SCR 1-24-49 |
| 299 | D 3-1-16, OOS 1951 and SCR 1-23-52 |

### Fleet Statistics: Peter Witt, 300–329

BLDR: G. C. Kuhlman; LEN: 51' 1"; W: 45,560 lbs; seats: 55; TRKS: Brill 51-E-1; MOT: 4WH340; CTL: K35G; HDL: Dayton 1583A; FEN: Eclipse; roof: arch; CR T 23.

| | |
|---|---|
| 300 | D 3-7-16, OOS 1949 and SCR 1-18-49 |
| 301 | G, D 3-8-16, OOS 1949 and SCR 2-8-49 |
| 302 | D 3-9-16, OOS E. 55th 1950 and SCR 8-10-51 |
| 303 | G, D 3-9-16, OOS 1949 and SCR 2-10-49 |
| 304 | G, D 3-1916, OOS 1949 and SCR 1-11-49 |
| 305 | G, D 3-10-16, OOS 1948 and SCR 6-4-48 |
| 306 | D 3-10-16, OOS 1949 and SCR 2-17-49 |
| 307 | D 3-11-16, OOS E. 55th 1950 and SCR 2-13-52 |
| 308 | D 3-11-16, OOS 1946? and SCR 7-10-47 |
| 309 | D 3-11-16, OOS 1950 and SCR 11-3-50 |
| 310 | D 3-14-16, OOS E. 55th 1950 and SCR 8-15-51 |
| 311 | D 3-16-16, OOS 1949 and SCR 9-12-49 |
| 312 | G, D 3-16-16, OOS 1949 and SCR 1-11-49 |
| 313 | D 3-17-16, OOS E. 55th 1950 and SCR 8-14-51 |
| 314 | D 3-18-16, OOS 1949 and SCR 3-1-49 |
| 315 | G, D 3-18-16, OOS 1946? and SCR 8-27-47 |
| 316 | D 3-20-16, ACC hit gasoline tanker, E. 105 and Carnegie, 17 injured, 1-29-47, SCR 8-27-47 |
| 317 | D 3-21-16, OOS E. 55th 1950 and SCR 2-13-52 |
| 318 | D 3-27-16, OOS E. 55th 1948 and SCR 12-9-49 |
| 319 | D 3-29-16, OOS 1948 and SCR 11-2-48 |
| 320 | G, D 4-1-16, OOS 1949 and SCR 2-16-49 |
| 321 | D 4-3-16, OOS 1949 and SCR 1-13-49 |
| 322 | G, D 4-6-16, OOS 1948 and SCR 6-23-48 |

Car 320, Peter Witt, G. C. Kuhlman, 1916. *(Cleveland Railway photo, Blaine Hays Collection)*

323     D 4-6-16, OOS 1949 and SCR 1-28-49
324     D 4-7-16, OOS E. 55th 1950 and SCR 2-6-52
325     D 4-11-16, OOS E. 55th 1950 and SCR 8-28-51
326     D 4-6-16, OOS 1951 and SCR 10-15-51
327     D 4-15-16, OOS E. 55th 1950 and SCR 3-13-52
328     G, D 4-26-16, OOS E. 55th 1948 and SCR 11-2-49
329     D 4-26-16, OOS 1949 and SCR 2-10-49

### Fleet Statistics: Peter Witt, 450–499

BLDR: G. C. Kuhlman; roof: arch; LEN: 52' 5"; W: 43,800 lbs; seats: 56; TRKS: 450–483, Brill 68 E-1; 484–499, Brill 51 E-1; MOT: 4WH340; CTL: K35G; headlight: Dayton 1583A; FEN: 450–476, Eclipse type B; 477–499, Eclipse; CR T 26.

450     D 7-26-20, seated 55, OOS 1950 and SCR 4-26-51
451     D 7-26-20, OOS E. 55th 1950 and SCR 8-28-51

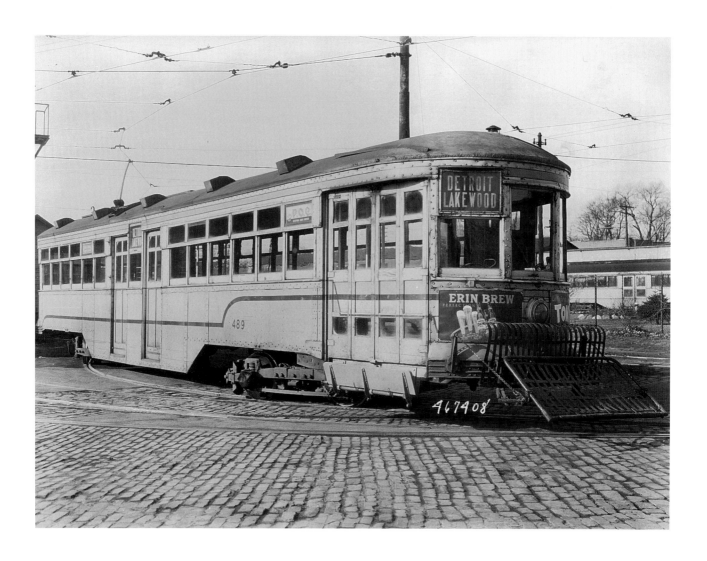

Car 489, Peter Witt, G. C. Kuhlman, 1920, CR type 26. *(Cleveland Railway photo, Blaine Hays Collection)*

| | |
|---|---|
| 452 | D 7-1920, OOS St. Clair 1949 and SCR 8-14-51 |
| 453 | D 7-29-20, seated 55, OOS St. Clair 1949 and SCR 8-6-51 |
| 454 | D 7-29-20, OOS 1949 and SCR 11-14-49 |
| 455 | D 8-6-20, OOS E. 55th 1950 and SCR 1-1951 |
| 456 | D 8-7-20, OOS St. Clair 1949 and SCR 8-15-51 |
| 457 | D 8-6-20, seated 55, OOS St. Clair 1950 and SCR 6-22-51 |
| 458 | D 8-7-20, OOS Brooklyn 1950 and SCR 9-25-51 |
| 459 | D 8-7-20, OOS E. 55th 1950 and SCR 9-25-51 |
| 460 | D 8-10-20, OOS 1950 and SCR 10-9-50 |
| 461 | D 8-11-20, OOS 1948 and SCR 5-28-48 |
| 462 | D 8-11-20, 51' 5" LEN, OOS St. Clair 1949 and SCR 6-22-51 |
| 463 | D 8-10-20, 51' 5" LEN, OOS 1949 and SCR 11-10-49 |
| 464 | D 8-13-20, OOS E. 55th 1950 and SCR 6-22-51 |
| 465 | G, D 8-13-20, OOS 1950 and SCR 1951 |
| 466 | G, D 8-13-20, OOS 1950 and SCR 4-10-51 |
| 467 | D 8-13-20, OOS 1950 and SCR 1-22-51 |
| 468 | D 8-17-20, seated 55, OOS E. 55th 1950 and SCR 9-25-51 |

469 D 8-18-20, 51' 5" LEN, OOS 1950 and SCR 6-29-51
470 D 8-20-20, 51' 5" LEN, OOS E. 55th 1948 and SCR 9-18-51
471 D 8-23-20, seated 55, OOS E. 55th 1948 and SCR 11-8-49
472 G, D 8-24-20, OOS E. 55th 1948 and SCR 9-6-50
473 D 8-27-20, destination sign over center front window, OOS E. 55th 1948, SCR 1951
474 D 8-25-20, OOS E. 55th 1948 and SCR 11-9-49
475 D 8-31-20, OOS E. 55th 1950 and SCR 9-10-51
476 G, D 8-27-20, OOS 1950 and SCR 3-13-51
477 D 9-2-20, OOS E. 55th 1950 and SCR 8-27-51
478 D 8-31-20, OOS 1949 and SCR 11-29-49
479 D 9-3-20, OOS E. 55th 1950 and SCR 9-10-51
480 G, D 9-8-20, OOS 1950 and SCR 10-25-50
481 D 9-2-20, OOS 1949 and SCR 11-9-49
482 G, D 8-31-20, OOS 1950 and SCR 9-15-50
483 G, D 9-9-20, OOS Brooklyn 1950 and SCR 8-31-51
484 D 9-9-20, OOS E. 55th 1950 and SCR 1951
485 D 9-14-20, OOS 1949 and SCR 11-9-49

Car 473, Peter Witt, G. C. Kuhlman, 1920, CR type 26—the only car in the series with the destination sign in the middle. *(Cleveland Railway photo, Blaine Hays Collection)*

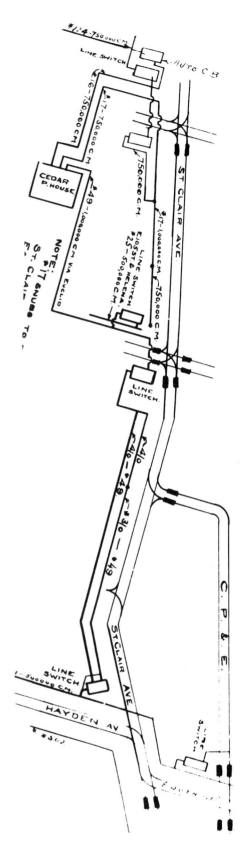

| | |
|---|---|
| 486 | D 9-8-20, OOS 1949 and SCR 11-29-49 |
| 487 | G, D 9-11-20, OOS 1950 and SCR 1951 |
| 488 | D 9-11-20, OOS E. 55th 1950 and SCR 1951 |
| 489 | G, D 9-22-20, seated 55, OOS E. 55th 1950 and SCR 8-31-51 |
| 490 | G, D 9-21-20, OOS 1948 and SCR 6-23-48 |
| 491 | G, D 9-20-20, OOS 1949 and SCR 1-3-49 |
| 492 | D 9-30-20, OOS 1950 and SCR 9-29-50 |
| 493 | D 9-30-20, OOS E. 55th 1948 and SCR 11-14-49 |
| 494 | D 9-30-20, OOS 1950 and SCR 7-21-50 |
| 495 | D 10-1-20, OOS 1950 and SCR 10-16-50 |
| 496 | D 10-3-20, OOS E. 55th 1950 and SCR 9-18-51 |
| 497 | 10-3-20, OOS 1950 and SCR 7-17-50 |
| 498 | G, D 10-3-20, OOS 1950 and SCR 9-15-50 |
| 499 | G, D 10-6-20, OOS 1950 and SCR 7-18-50 |

### Fleet Statistics: Peter Witt, 350 and 351–449

BLDR: G. C. Kuhlman; roof: arch; LEN: 52' 5"; W: 43,800 lbs; seats: 56; TRKS: Brill 68 E-1; MOT: car 350, GE (later changed to WH); 351–449, 4WH340; CTL: K35G; HDL: Dayton 1583A; FEN: Eclipse; 350, CR T 24; 351–449, CR T 25.

| | |
|---|---|
| 350 | D 7-7-23, OOS 1949 and SCR 10-25-49 |
| 351 | G, D 7-7-23, OOS 1950 and SCR 10-10-50 |
| 352 | G, D 7-11-23, OOS St. Clair 1950 and SCR 8-6-51 |
| 353 | G, D 7-12-23, OOS St. Clair 1950 and SCR 8-21-51 |
| 354 | G, D 7-16-23, OOS 1950 and SCR probably 1950? |
| 355 | G, D 7-13-23, OOS 1948 and SCR 9-10-48 |
| 356 | D 7-14-23, OOS St. Clair 1950 and SCR 9-17-51 |
| 357 | G, D 7-16-23, OOS St. Clair 1950 and SCR 1-14-52 |
| 358 | D 7-17-23, OOS 1950 and SCR 8-18-50 |
| 359 | G, D 7-18-23, rubber encased silent TRKS, OOS St. Clair 1950, SCR 1-14-52 |
| 360 | D 7-18-23, OOS St. Clair 1950 and SCR 11-21-51 |
| 361 | D 7-19-23, OOS St. Clair 1950 and SCR 1-14-52 |
| 362 | G, D 7-20-23, OOS 1949 and SCR 1-6-49 |
| 363 | G, D 7-21-23, OOS 1950, CON to line car 0363 with addition of roof platform and side stairway and removal of seats, operated in "Parade of Progress" 4-27-52, SCR 3-30-54 |
| 364 | G, D 7-21-23, OOS 1949 and SCR 10-21-49 |
| 365 | G, D 7-24-23, OOS 1950 and SCR 1-22-51 |
| 366 | G, D 7-25-23, OOS St. Clair 1949 and SCR 10-20-50 |
| 367 | D 7-26-23, OOS St. Clair 1949 and SCR probably 1950? |
| 368 | G, D 7-27-23, OOS St. Clair 1950 and SCR 8-23-51 |
| 369 | D 7-28-23, OOS 1948 and SCR 6-15-48 |

A publicity photo of a trailer train with motor 415 and trailers 2391, 2388, 2392 (more than one trailer was not used in actual service). *(Cleveland Railway photo, Blaine Hays Collection)*

| | |
|---|---|
| 370 | D 7-31-23, OOS 1950 and SCR 12-6-50 |
| 371 | D 8-1-23, had direct air brakes, OOS St. Clair in 1950, SCR 8-13-51 |
| 372 | D 8-3-23, OOS St. Clair 1949 and SCR 8-23-51 |
| 373 | D 8-4-23, OOS E. 55th 1950 and SCR 11-21-51 |
| 374 | D 8-7-23, OOS 1949 and SCR 11-2-49 (listed in record as SCR 10-31-49) |
| 375 | D 8-8-23, OOS 1950 and SCR 10-16-50 |
| 376 | D 8-10-23, OOS 1950 and SCR 10-9-50 |
| 377 | D 8-1-23, OOS E. 55th 1950 and SCR 9-17-51 |
| 378 | G, D 8-15-23, OOS E. 55th 1950 and SCR 1-1951 |
| 379 | G, D 8-15-23, OOS Harvard 1948 and SCR 10-27-49 |
| 380 | D 8-17-23, OOS St. Clair 1949 and SCR 10-20-50 |
| 381 | G, D 8-18-23, OOS St. Clair 1950 and SCR 10-8-51 |
| 382 | G, D 8-21-23, OOS 1950 and SCR 11-3-50 |
| 383 | G, D 8-22-23, OOS St. Clair 1949 and SCR 10-25-50 |
| 384 | D 8-23-23, OOS St. Clair 1949 and SCR 9-25-51 |
| 385 | G, D 8-24-23, OOS 1949 and SCR 10-25-49 |
| 386 | D 8-25-23, OOS 1950 and SCR 3-27-51 |
| 387 | G, D 8-28-23, OOS E. 55th 1950 and SCR 9-25-51 |
| 388 | G, D 8-28-23, OOS St. Clair 1950 and SCR 1-3-52 |
| 389 | G, D 8-29-23, OOS St. Clair 1950 and SCR 8-17-51 |

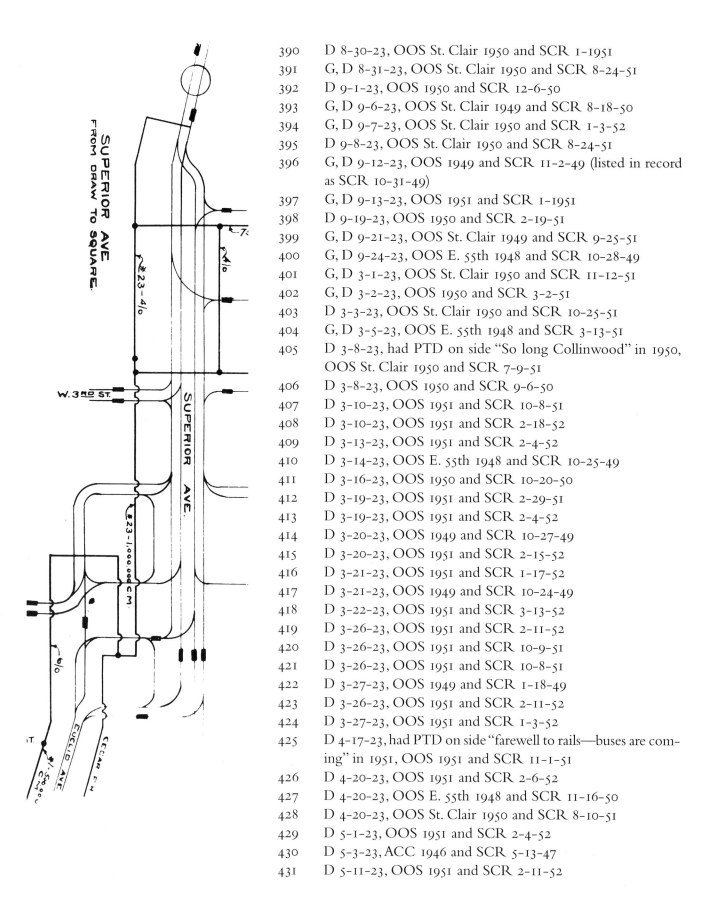

| | |
|---|---|
| 390 | D 8-30-23, OOS St. Clair 1950 and SCR 1-1951 |
| 391 | G, D 8-31-23, OOS St. Clair 1950 and SCR 8-24-51 |
| 392 | D 9-1-23, OOS 1950 and SCR 12-6-50 |
| 393 | G, D 9-6-23, OOS St. Clair 1949 and SCR 8-18-50 |
| 394 | G, D 9-7-23, OOS St. Clair 1950 and SCR 1-3-52 |
| 395 | D 9-8-23, OOS St. Clair 1950 and SCR 8-24-51 |
| 396 | G, D 9-12-23, OOS 1949 and SCR 11-2-49 (listed in record as SCR 10-31-49) |
| 397 | G, D 9-13-23, OOS 1951 and SCR 1-1951 |
| 398 | D 9-19-23, OOS 1950 and SCR 2-19-51 |
| 399 | G, D 9-21-23, OOS St. Clair 1949 and SCR 9-25-51 |
| 400 | G, D 9-24-23, OOS E. 55th 1948 and SCR 10-28-49 |
| 401 | G, D 3-1-23, OOS St. Clair 1950 and SCR 11-12-51 |
| 402 | G, D 3-2-23, OOS 1950 and SCR 3-2-51 |
| 403 | D 3-3-23, OOS St. Clair 1950 and SCR 10-25-51 |
| 404 | G, D 3-5-23, OOS E. 55th 1948 and SCR 3-13-51 |
| 405 | D 3-8-23, had PTD on side "So long Collinwood" in 1950, OOS St. Clair 1950 and SCR 7-9-51 |
| 406 | D 3-8-23, OOS 1950 and SCR 9-6-50 |
| 407 | D 3-10-23, OOS 1951 and SCR 10-8-51 |
| 408 | D 3-10-23, OOS 1951 and SCR 2-18-52 |
| 409 | D 3-13-23, OOS 1951 and SCR 2-4-52 |
| 410 | D 3-14-23, OOS E. 55th 1948 and SCR 10-25-49 |
| 411 | D 3-16-23, OOS 1950 and SCR 10-20-50 |
| 412 | D 3-19-23, OOS 1951 and SCR 2-29-51 |
| 413 | D 3-19-23, OOS 1951 and SCR 2-4-52 |
| 414 | D 3-20-23, OOS 1949 and SCR 10-27-49 |
| 415 | D 3-20-23, OOS 1951 and SCR 2-15-52 |
| 416 | D 3-21-23, OOS 1951 and SCR 1-17-52 |
| 417 | D 3-21-23, OOS 1949 and SCR 10-24-49 |
| 418 | D 3-22-23, OOS 1951 and SCR 3-13-52 |
| 419 | D 3-26-23, OOS 1951 and SCR 2-11-52 |
| 420 | D 3-26-23, OOS 1951 and SCR 10-9-51 |
| 421 | D 3-26-23, OOS 1951 and SCR 10-8-51 |
| 422 | D 3-27-23, OOS 1949 and SCR 1-18-49 |
| 423 | D 3-26-23, OOS 1951 and SCR 2-11-52 |
| 424 | D 3-27-23, OOS 1951 and SCR 1-3-52 |
| 425 | D 4-17-23, had PTD on side "farewell to rails—buses are coming" in 1951, OOS 1951 and SCR 11-1-51 |
| 426 | D 4-20-23, OOS 1951 and SCR 2-6-52 |
| 427 | D 4-20-23, OOS E. 55th 1948 and SCR 11-16-50 |
| 428 | D 4-20-23, OOS St. Clair 1950 and SCR 8-10-51 |
| 429 | D 5-1-23, OOS 1951 and SCR 2-4-52 |
| 430 | D 5-3-23, ACC 1946 and SCR 5-13-47 |
| 431 | D 5-11-23, OOS 1951 and SCR 2-11-52 |

432  D 5-12-23, OOS St. Clair 1950 and SCR 1-1951
433  D 5-15-23, OOS 1951 and SCR 2-8-52
434  D 5-18-23, OOS 1950 and SCR 10-25-50
435  D 5-23-23, OOS St. Clair 1950 and SCR 2-8-52
436  D 5-25-23, OOS E. 55th 1950 and SCR 9-17-51
437  D 5-26-23, OOS 1950 and SCR 11-3-50
438  D 6-2-23, OOS St. Clair 1950, in 1951 had PTD on either side "Goodbye to Rails," and SCR 10-8-51
439  D 6-1-23, OOS St. Clair 1950 and SCR 2-8-52
440  D 5-29-23, OOS E. 55th 1950 and SCR 9-17-51
441  D 6-7-23, OOS 1949 and SCR 11-2-49
442  D 6-9-23, OOS 1950 and SCR 9-29-50
443  D 6-13-23, OOS St. Clair 1950 and SCR 9-25-51
444  D 6-22 23, OOS St. Clair 1950 and SCR 2-15-52
445  D 6-19-23, OOS 1951 and SCR 2-6-52
446  D 6-20-23, OOS St. Clair 1950 and SCR 11-12-51
447  D 6-25-23, OOS 1948 and SCR 7-26-48
448  D 6-28-23, OOS 1950 and SCR 6-29-51
449  D 7-2-23, OOS 1950 and SCR 3-2-51

### *Fleet Statistics: Peter Witt, 150–174*

BLDR: Cleveland Railway; roof: arch; W: 41,600 lbs; seats: 56; TRKS: 150–164, Brill 51-E-1 (swapped with CIRR cars in 1920s); 165–174, Brill 68-E-1; MOT: 4WH340; CTL: K35G; HDL: Dayton 1583A; FEN: Eclipse; LEN: 151–157 and 161–174, 51' 2"; CR T 27.

150  FN 1301, D 9-16-24, LEN 51' 4", OOS in 1951 and SCR 6-23-52
151  G, D 1-7-25, OOS in 1951 and SCR 6-13-52
152  G, D 1-1925, OOS in 1951 and SCR 6-4-52
153  G, D 1-1925, Eclipse wheel guards as a test, OOS in 1951 and SCR 6-4-52
154  G, D 1-1925, OOS in 1951 and SCR 6-23-52
155  G, D 1-1925, OOS in 1951 and SCR 6-23-52
156  G, D 1-1925, OOS in 1951 and SCR 6-13-52
157  G, D 1-1925, OOS in 1951 and SCR 6-13-52
158  G, D 1-1925, LEN 51' 1", OOS in 1951 and SCR 1-31-52
159  G, D 1-1925, LEN 51' 1", OOS in 1951 and SCR 6-4-52
160  D 2-1925, LEN 51' 1", OOS in 1951 and SCR 6-4-52
161  G, D 2-1925, OOS in 1951 and SCR 6-13-52
162  D 2-1925, OOS in 1951 and SCR 7-10-52
163  G, D 2-1925, OOS in 1951 and SCR 6-23-52
164  G, D 2-1925, OOS in 1951 and SCR 6-23-52
165  G, D 2-1925, W 43,200 lbs, OOS in 1951 and SCR 6-23-52

Car 151, Peter Witt, Cleveland Railway, 1925, CR type 27. *(Cleveland Transit photo, Blaine Hays Collection)*

| | |
|---|---|
| 166 | G, D 2-1925, W 43,200 lbs, CON OM 10-24-35, OOS in 1951 and SCR 6-4-52 |
| 167 | G, D 3-1925, W 43,200 lbs, CON OM 10-22-35, OOS in 1951 and SCR 6-23-52 |
| 168 | G, D 3-1925, W 43,200 lbs, CON OM 10-28-35, OOS in 1951 and SCR 6-4-52 |
| 169 | D 3-1925, W 43,200 lbs, CON OM 10-19-35, OOS in 1951 and SCR 6-23-52 |
| 170 | D 3-1925, W 43,200 lbs, CON OM 10-23-35, ACC 3-22-41 on Kinsman, OOS in 1951 and SCR 6-23-52 |
| 171 | D 3-1925, W 43,200 lbs, CON OM 10-18-35, wrecked March 1951 and SCR 6-23-52 |
| 172 | G, D 3-1925, W 42,600 lbs, CON OM 10-18-35, OOS in 1951 and SCR 6-23-52 |
| 173 | G, D 3-1925, W 43,200 lbs, CON OM 10-21-35, OOS in 1951 and SCR 6-13-52 |
| 174 | G, D 3-1925, W 43,200 lbs, CON OM 10-12-35, OOS in 1951 and SCR 6-23-52 |

## Fleet Statistics: Peter Witt, 1301–1376

BLDR: CR; doors: 1311–1339, 1-man air type 1301–1310 and 1340–1376, fold hand operated; seats: 1311–1339, 53; 1340–1376, 55; HDL: Dayton 1583A; MOT: 1301–1348, 4WH340; 1349–1375, 4WH340P; 1376, 4WH510E; signs: two Hunter; TRKS: 1301–1310, Standard C-50; 1311–1375, Brill 68E-1; 1376, Brill 68E-1; aluminum; LEN: 51' 2"; W: 1301–1310, 41,740 lbs; 1311–1375, 42,600 lbs; 1376, 30,500 lbs; CTL: K35G; 1322–1339 had backup equipment; cost: $10,417 each; crew size: 1301–1310 and 1340–1376, two; 1311–1339, one; CR T 1301–1372 T 28; 1373–5 T 29; 1376 T 30. The 1300s were the newest cars at the time of the 1926 American Electric Railway Association (AERA) convention, and many experiments were tried with them. Car 1373 with trailer 1995 was on display at the 1927 Industrial Exposition, and these CR-built cars were given much play when they first arrived.

1301    G, D 4-1925, OOS 1951, SCR 6-23-52
1302    G, D 1925, OOS E. 55th 1951, SCR 7-14-52

Car 1325, Peter Witt, Cleveland Railway, 1925, CR type 28. *(Cleveland Railway photo, Blaine Hays Collection)*

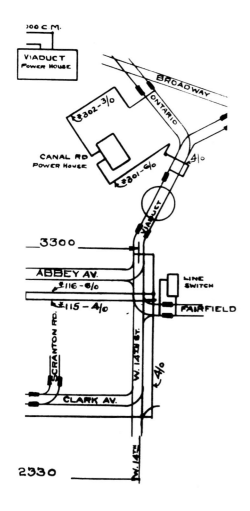

| | |
|---|---|
| 1303 | G, D 4-1925, OOS E. 55th 1951, SCR 7-10-52 |
| 1304 | G, D 4-1925, OOS 1951, SCR 7-14-52 |
| 1305 | G, D 4-1925, OOS 1951, SCR 8-18-52 |
| 1306 | G, D 4-1925, OOS Brooklyn 1950, CON to storage at Harvard 1953, SCR 9-4-53 |
| 1307 | G, D 4-1925, OOS Brooklyn 1950, SCR 12-5-51 |
| 1308 | G, D 4-1925, OOS E. 55th 1950, SCR 1-31-52 |
| 1309 | G, D 4-1925, OOS 1951, CON to storage at Harvard 1953, SCR 9-29-53 |
| 1310 | G, D 4-1925, OOS 1951, SCR 6-23-52 |
| 1311 | G, D 5-1925, OOS 1951, SCR 8-22-52 |
| 1312 | G, D 5-1925, OOS 1951, SCR 8-12-52 |
| 1313 | G, D 5-1925, OOS 1951, SCR 7-3-52 |
| 1314 | G, D 5-1925, OOS 1951, SCR 1952 |
| 1315 | G, D 5-1925, OOS 1951, SCR 1952 |
| 1316 | G, D 5-1925, OOS 1951, SCR 7-30-52 |
| 1317 | G, D 5-1925, ACC W. 65th and Herman on 4-27-46, OOS 1951, SCR 7-3-52 |
| 1318 | D 5-1925, OOS 1951, SCR 8-12-52 |
| 1319 | G, D 5-1925, OOS 1951, SCR 8-12-52 |
| 1320 | D 6-1925, OOS 1951, SCR 7-10-52 |
| 1321 | G, D 6-1925, OOS 1951, SCR 7-30-52 |
| 1322 | D 6-1925, OOS 1951, SCR 8-12-52 |
| 1323 | D 6-1925, OOS 1951, SCR 8-12-52 |
| 1324 | D 6-1925, OOS 1951, SCR 7-14-52 |
| 1325 | G, D 6-1925, OOS 1951, SCR 7-30-52 |
| 1326 | D 6-1925, OOS E. 55th 1951, SCR 7-10-52 |
| 1327 | G, D 8-1925, had serious front-end ACC in 1948, OOS 1951, SCR 10-15-51 |
| 1328 | G, D 8-1925, OOS 1951, SCR 1952 |
| 1329 | G, D 8-1925, OOS 1951, SCR 7-14-52 |
| 1330 | G, D 8-1925, OOS 1951, SCR 8-13-52 |
| 1331 | G, D 8-1925, OOS 1951, SCR 8-12-52 |
| 1332 | G, D 8-1925, OOS 1951, SCR 8-13-52 |
| 1333 | G, D 8-1925, OOS 1951, SCR 8-19-52 |
| 1334 | G, D 8-1925, OOS 1951, SCR 8-18-52 |
| 1335 | G, D 8-1925, OOS 1951, SCR 1952 |
| 1336 | G, D 8-1925, OOS 1951, SCR 7-30-52 |
| 1337 | G, D 8-1925, OOS 1951, SCR 7-10-52 |
| 1338 | G, D 8-1925, OOS 1951, SCR 1-31-52 |
| 1339 | G, D 8-1925, ACQ TRKS from 449-68E-1 on 6-28-50, OOS 1951, SCR 1952 |
| 1340 | G, D 8-1925, OOS 1951, SCR 6-23-52 |
| 1341 | G, D 8-1925, OOS 1951, SCR 8-19-52 |
| 1342 | G, D 9-1925, OOS 1951, SCR 1952 |
| 1343 | G, D 9-1925, OOS 1951, SCR 8-19-52 |

1344　D 9-1925, OOS 1951, SCR 1952
1345　G, D 9-1925, OOS 1951, SCR 6-23-52
1346　G, D 9-16-25, Timken roller-bearing TRKS, OOS 1951, SCR 2-29-51
1347　G, D 9-1925, Timken roller-bearing TRKS, OOS 1951, SCR 8-18-52
1348　G, D 9-1925, Timken roller-bearing TRKS, OOS 1951, SCR 8-18-52
1349　G, D 9-1925, Timken roller-bearing TRKS, OOS 1951, SCR 8-18-52
1350　G, D 9-1925, experimental roof ventilators, OOS 1951, SCR 3-18-52
1351　G, D 10-1925, OOS 1951, SCR 2-26-52
1352　G, D 10-1925, OOS E. 55th 1951, SCR 7-14-52
1353　G, D 10-1925, OOS 1951, SCR 8-18-53
1354　G, D 10-1925, OOS E. 55th 1951, SCR 7-14-52
1355　G, D 10-1925, OOS 1951, SCR 2-21-52
1356　G, D 10-1925, OOS 1951, SCR 2-21-52
1357　G, D 10-26-25, OOS 1951, SCR 2-26-52
1358　G, D 10-1925, OOS 1951, SCR 2-21-52
1359　G, D 10-1925, OOS 1951, SCR 2-22-52
1360　G, D 11-1925, experimental roof ventilators, OOS 1951, SCR 2-29-52
1361　D 10-1925, rewound for higher speed, CON to storage bin at Harvard 1953, used as Harvard Shop scrapping tow car, SCR 5-5-54 (last CTS-owned motorcar scrapped)
1362　G, D 10-1925, OOS 1951, SCR 12-5-51
1363　D 10-1925, OOS E. 55th 1950, SCR 12-5-51
1364　G, 10-1925, OOS E. 55th 1951, SCR 7-10-52
1365　G, D 11-1925, OOS 1951, SCR 8-13-52
1366　D 11-1925, OOS 1951, SCR 2-22-52
1367　D 11-1925, OOS E. 55th 1951, SCR 6-23-52
1368　G, D 11-1925, OOS 1951, SCR 2-21-52
1369　G, D 11-1925, OOS 1951, SCR 2-21-52
1370　G, D 11-1925, experimental roof ventilators, OOS 1951, SCR 1-17-52
1371　G, D 11-1925, experimental roof ventilators, OOS 1951, SCR 2-29-52
1372　D 11-1925, experimental roof ventilators, OOS 1951, SCR 2-21-52
1373　G, D 11-1925, experimental roof ventilators, leather seats, OOS 1951, SCR 2-26-52
1374　D 11-1925, experimental roof ventilators, leather seats, OOS E. 55th 1951, SCR 7-14-52
1375　D 11-1925, experimental roof ventilators, leather seats, OOS 1951, SCR 7-10-52

1376    D 10-4 to the stage of Public Hall where it was displayed to 10-8-26 during AERA convention; experimental roof ventilators, CTL K35KK; BLT entirely of aluminum; including body, farebox, TRKS, coupler, Eclipse FEN and HDL, cost $18,000; placed in revenue service 12-1-26, OOS 1951, SCR 9-3-52

### Fleet Statistics: Articulated Peter Witt, 5000–5027

BLDR: G. C. Kuhlman Car Co.; LEN: 101.2'; seats: front, 44; rear, 56; HDL: Golden Glow; TRKS: three Brill 68E-2; MOT: six WH340P; crew size: three; FEN: wheel guard, Eclipse B; W: 80,400 lbs; cost: $30,351 each; CTL: HL control 15-B-20; CR T 31. All cars OOS in 1952 (except as noted) and sold to the A. Shaw Co. and SCR en masse at Harvard on 3-24-53; a great effort was made to sell these cars in South America.

Car 5008, Peter Witt articulated, G. C. Kuhlman 1928, CR type 31. (*Jim Spangler photo*)

5000    D 10-24-27, had Master 15-B-20 CTL
5001    G, D 7-27-28
5002    G, D 7-27-28, OOS 1951

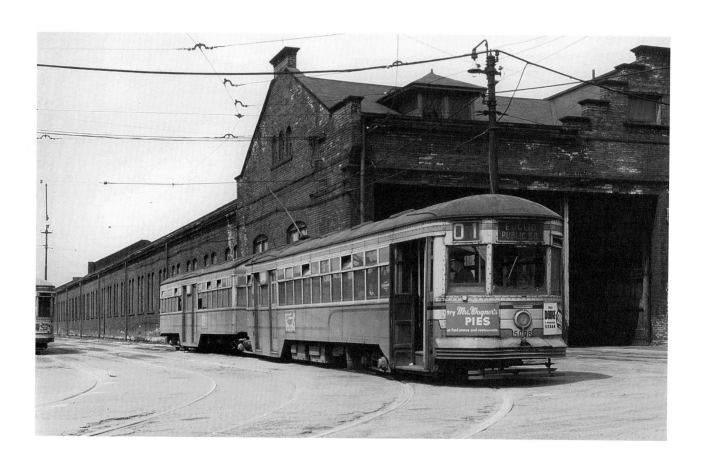

Car 5000, inside the G. C. Kuhlman plant in its original configuration. *(Cleveland Railway photo, Blaine Hays Collection)*

| | |
|---|---|
| 5003 | G, D 7-29-28, OOS 1951 |
| 5004 | G, D 7-29-28 |
| 5005 | G, D 8-1-28 |
| 5006 | G, D 8-1-28 |
| 5007 | G, D 8-4-28 |
| 5008 | G, D 8-7-28 |
| 5009 | G, D 8-9-28 |
| 5010 | G, D 8-10-28 |
| 5011 | G, D 8-14-28, OOS 1951 |
| 5012 | G, D 8-15-28 |
| 5013 | G, D 8-17-28, OOS 1951 |
| 5014 | D 8-21-28, equipped with 6 WH 341 P-A-1 MOT and pantograph installed for test run on Shaker Rapid in 1931 during Van Sweringen control; car unable to clear loop under Terminal Tower; tests suspended, OOS 1951 |
| 5015 | G, D 8-23-28 |
| 5016 | G, D 8-26-28 |
| 5017 | G, D 9-8-28, equipped with Master 15-B-20 CTL |
| 5018 | G, D 9-11-28 |
| 5019 | G, D 9-13-28 |
| 5020 | G, D 9-19-28 |
| 5021 | G, D 9-21-28 |
| 5022 | G, D 9-25-28 |
| 5023 | G, D 9-27-28 |

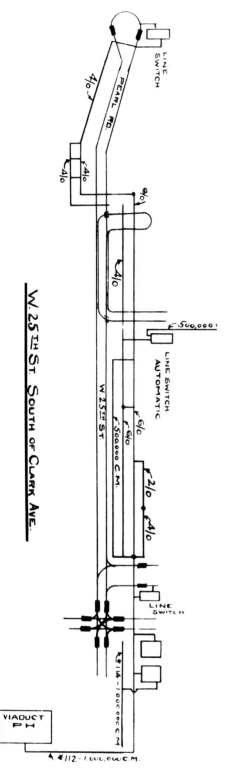

5024   G, D 9-29-28, equipped with Master 15-B-20 CTL
5025   G, D 4-13-29, last 5000 to operate, in "Parade of Progress" 4-27-52
5026   G, D 4-20-29
5027   G, D 5-8-29

### Fleet Statistics: Peter Witt, 4000–4149

BLDR: G. C. Kuhlman Car Co.; LEN: 4000–4045, 53' 7"; 4046–4149, 53' 6"; seats: 55; HDL: Golden Glow; FEN: Eclipse wheel guard, type B; W: 4000–4049, 48,000 lbs; 4050–4149, 48,200 lbs; cost each: $18,050; TRKS: Brill 68E-2; crew size: 4000–4049, one or two; 4050–4149, two; backup control; MOT: 4WH340P; CTL: 4000–4051, HL master 15-B-20; 4052–4149, HL control 15-B-20; some cars had backup CTL (designated BU); CR T 32. Articulated car 5000 was completed by Kuhlman as the first of the "Ultra Peter Witts" on 10-24-27. A single-unit car of the same design was finished 12-27-27 and numbered 333. Both of these "first of fleets" had three front windows of the same size and a split destination sign. After testing by CR, a wider front window was requested for better visibility and elimination of the split destination sign. The company changed the number series to the 4000s for single-unit cars and kept the 5000 numbers for articulateds. Car 5001 was completed, incorporating the changes, on 7-27-28, and car 4001 followed on 9-29-28. Car 333 was REB and REN 4000 on 1-9-29. In 1939 CR began renovating this type car with new interior lighting and rubber flooring; eight cars received more extensive renovation. On their last day, Superior cars carried painted sides that said "Let me rust in peace" and "so long folks, just bury my tracks."

4000   G, D 12-27-27, FN 333, REB 1-9-29, SCR 7-14-53
4001   G, D 9-29-28, OOS St. Clair 1951, SCR 7-14-53
4002   D 10-4-28, SCR 7-14-53
4003   G, D 10-4-28, SCR 7-14-53
4004   G, D 10-5-28, dubbed the CTS Jinx Car by shop crew for having 20 accidents in a two-year stretch; OOS ACC stored at Denison in 1951; SCR 7-14-53
4005   G, D 10-8-28, SCR 7-14-53
4006   G, D 10-9-28, OOS St. Clair 1951, SCR 7-14-53
4007   G, D 10-12-28, OOS ACC stored at Denison in 1951, SCR 7-14-53
4008   G, D 10-18-28, SCR 7-14-53
4009   G, D 10-18-28, OOS St. Clair 1951, SCR 7-14-53
4010   D 10-27-28, SCR 7-14-53
4011   G, D 10-25-28, SCR 7-14-53

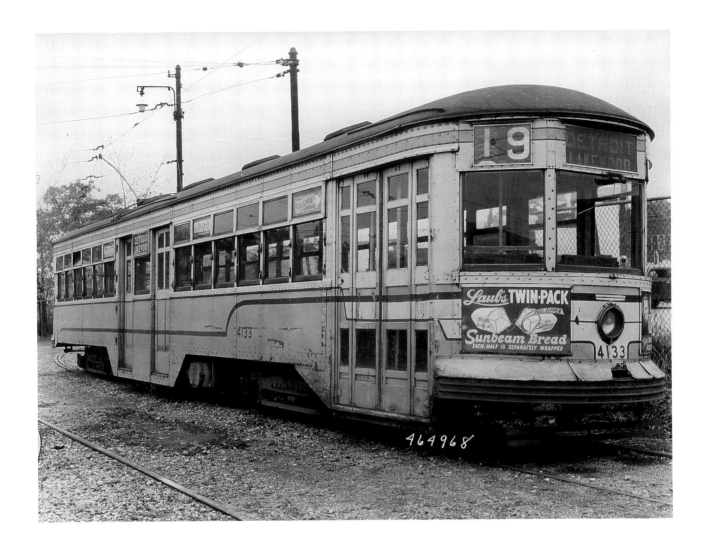

Car 4133, Peter Witt, G. C. Kuhlman, 1930, CR type 32. *(Cleveland Transit photo, Blaine Hays Collection)*

4012 G, D 10-19-28, CON to storage car at Harvard 1951, SCR 4-15-53
4013 G, D 10-19-28, SCR 7-14-53
4014 G, D 10-23-28, SCR 7-14-53
4015 G, D 10-23-28, SCR 7-14-53
4016 D 10-20-28, SCR 7-14-53
4017 D 10-20-28, SCR 7-14-53
4018 G, D 10-25-28, SCR 7-14-53
4019 G, D 10-27-28, ACC High Level Bridge on 9-4-43, SCR 7-14-53
4020 D 10-27-28, SCR 7-14-53
4021 G, D 10-30-28, SCR 7-14-53
4022 G, D 10-30-28, SCR 7-14-53
4023 G, D 10-31-28, painted for "Green Cross for Safety" campaign in 1939, SCR 10-13-53
4024 G, D 10-31-28, SCR 7-14-53

| | |
|---|---|
| 4025 | G, D 11-2-28, SCR 7-14-53 |
| 4026 | G, D 11-2-28, REB same as car 4073 in 1939, SCR 7-14-53 |
| 4027 | G, D 11-2-28, SCR 7-14-53 |
| 4028 | G, D 11-2-28, SCR 7-14-53 |
| 4029 | G, D 11-7-28, involved in serious ACC in 1947, seats transferred to car 1376, SCR 12-9-47 |
| 4030 | D 11-7-28, SCR 7-14-53 |
| 4031 | G, D 11-9-28, SCR 7-14-53 |
| 4032 | D 11-9-28, ACC OOS St. Clair 1951, SCR 7-14-53 |
| 4033 | G, D 11-9-28, SCR 7-14-53 |
| 4034 | G, D 11-10-28, had vertical handrails throughout and was known as "Warsaw" because of its "forest" of poles, OOS 1951, SCR 7-18-52 |
| 4035 | D 11-13-28, SCR 7-14-53 |
| 4036 | D 11-17-28, SCR 7-14-53 |
| 4037 | D 11-13-28, OOS Denison 1951, SCR 7-14-53 |
| 4038 | D 11-17-28, SCR 7-14-53 |
| 4039 | D 11-20-28, SCR 7-14-53 |
| 4040 | D 11-20-28, SCR 7-14-53 |
| 4041 | D 11-23-28, SCR 7-14-53 |
| 4042 | D 11-23-28, SCR 7-14-53 |
| 4043 | D 11-27-28, SCR 7-14-53 |
| 4044 | D 11-27-28, SCR 7-14-53 |
| 4045 | D 12-18-28, SCR 7-14-53 |
| 4046 | D 11-27-28, SCR 7-14-53 |
| 4047 | D 12-2-28, SCR 7-14-53 |
| 4048 | D 12-13-28, rewired for train operation in 1945, SCR 7-14-53 |
| 4049 | D 12-18-28, rewired for train operation in 1945, had Eclipse FEN for a short time, SCR 7-14-53 |
| 4050 | G, D 12-4-29, BU, used on the Shaker Rapid in 1935, SCR 10-29-53 |
| 4051 | G, D 12-3-29, BU, used on the Shaker Rapid in 1935, last revenue car on both Clark and Madison lines, SCR 3-30-54 |
| 4052 | D 11-25-29, BU, SCR 11-10-53 |
| 4053 | D 11-27-29, SCR 11-10-53 |
| 4054 | D 12-9-29, BU, SCR 11-10-53 |
| 4055 | G, D 12-7-29, BU, OOS 1951, SCR 7-10-52 |
| 4056 | G, D 12-6-29, BU, OOS 1951, SCR 11-10-53 |
| 4057 | D 12-9-29, SCR 3-30-54 |
| 4058 | D 12-10-29, SCR 11-10-53 |
| 4059 | D 12-13-29, SCR 10-29-53 |
| 4060 | D 12-13-29, in serious ACC at Cedar Glen on 1-19-41, SCR 10-29-53 |
| 4061 | D 12-11-29, SCR 7-14-53 |
| 4062 | D 12-12-29, BU, SCR 3-30-54 |
| 4063 | D 12-14-29, held for Shaker Rapid in 1953, not taken, SCR |

3-30-54

4064 D 12-17-29, special GE indirect lighting installed in 1935 consisting of 100 30-volt lamps and 5 101-watt lamps, dubbed the "Sight Saver" car; car involved in serious accident with 4060 at Cedar Glen on 1-19-41, SCR 10-29-53

4065 D 12-14-29, OOS 1951, SCR 7-10-52

4066 G, D 12-18-29, REB same as 4073 in 1939, first car with new Raymond Loewy orange/cream/tan paint on 1-7-41, ACC in 1951 OSS E. 55th, SCR 7-14-53

4067 D 12-23-29, OOS 1951, SCR 2-22-52

4068 G, D 12-23-29, BU, CON to OM 1941, stored, retrieved and rebuilt to two-man for wartime riding in 1942, officially designated as "experiment car" by CR, all trials conducted on this car, SCR 11-10-53

4069 D 12-26-29, OOS 1951, SCR 2-29-52

4070 G, D 12-23-29, SCR 10-29-53

4071 D 12-26-29, SCR probably 1953

4072 G, D 12-28-29, SCR 10-29-53

4073 G, D 12-26-29, 1939 car equipped with resilient wheels, rubber-mounted journal boxes, rubber bolster units replacing elliptic springs, and the underside insulated with hair felt and tarpaper; first car PTD in the Raymond Loewy green/gray scheme, SCR 10-29-53

4074 G, D 12-28-29, SCR 10-29-53

4075 D 12-31-29, SCR 10-29-53

4076 D 12-28-29, SCR 10-29-53

4077 D 12-30-29, SCR 10-29-53

4078 D 12-31-29, SCR 10-29-53

4079 G, D 1-3-30, SCR 10-29-53

4080 D 1-9-30, SCR 10-29-53

4081 G, D 1-11-30, BU, SCR 11-10-53

4082 G, D 1-11-30, BU, had flag-bunting for last day on Madison 1-24-54, SCR 3-30-54

4083 G, D 1-14-30, BU, had flag-bunting for last day on Madison 1-24-54, SCR 3-30-54

4084 D 1-14-30, SCR 10-29-53

4085 G, D 1-14-30, SCR 10-29-53

4086 G, D 1-18-30, SCR 10-29-53

4087 G, D 1-18-30, BU, SCR 11-10-53

4088 G, D 1-25-30, SCR 10-29-53

4089 D 1-22-30, ACC on Euclid at E. 37th St. head-on with bus, 7-26-46, SCR 10-29-53

4090 G, D 1-25-30, SCR 10-29-53

4091 G, D 1-23-30, BU, Fabreeka installed, SCR 7-14-53

4092 D 1-25-30, SCR 10-29-53

4093 D 1-28-30, SCR 1954

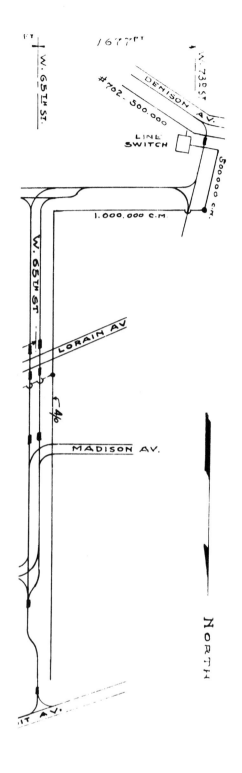

4094 G, D 1-28-30?, BU, had flag-bunting for last day on Madison 1-24-54, SCR 3-30-54

4095 G, D 1-28-30, PTD red, white, and blue for Marine recruiting in 1948, SCR 10-29-53

4096 D 1-31-30, SCR 10-29-53

4097 D 1-29-30, SCR 10-29-53

4098 G, D 2-1-30, SCR 11-10-53

4099 D 2-7-30, held for Shaker Rapid in 1953 but not taken, SCR 10-29-53

4100 G, D 2-7-30, BU, SCR 3-30-54

4101 G, D 2-5-30, held for Shaker Rapid in 1953 but not taken, SCR 10-29-53

4102 G, D 2-4-30, held for Shaker Rapid in 1953, not taken, SCR 10-29-53

4103 G, D 2-12-30, SCR 10-29-53

4104 G, D 2-13-30, BU, SCR 11-10-53

4105 G, D 2-13-30, BU, wired for higher speed, SCR 3-31-54

4106 G, D 2-11-30, BU, SCR 3-31-54

4107 G, D 2-10-30, equipped with two-notch dynamic brakes and two 2-unit line switches, REB same as 4073 in 1939, held for Shaker Rapid in 1953 but not taken, SCR 10-29-53

4108 G, D 2-15-30, BU, wired for higher speed, ACC 2-11-35, SCR 3-30-54

4109 G, D 2-15-30, BU, SCR 3-30-54

4110 G, D 2-14-30, BU, SCR 3-30-54

4111 G, D 2-18-30, BU, SCR 11-10-53

4112 G, D 2-19-30, had flag-bunting for last day on Madison 1-24-54, BU, SCR 3-30-54

4113 G, D 2-20-30, BU, PTD red, white, and blue for war bonds in 1943, SCR 11-10-53

4114 G, D 2-21-30, BU, SCR 11-10-53

4115 G, D 2-22-30, ACC at E. 9th and Euclid in 1937, SCR 3-30-54

4116 G, D 2-24-30, BU, SCR 11-10-53

4117 G, D 2-25-30, BU, four rubber bolsters installed, had flag-bunting for last day on Madison 1-24-54, SCR 3-30-54

4118 G, D 2-26-30, BU, SCR 11-10-53

4119 D 2-27-30, BU, SCR 3-30-54

4120 G, D 3-4-30, BU, Fabreeka installed and rubber bolsters, SCR 3-30-54

4121 G, D 3-5-30, BU, SCR 10-29-53

4122 G, D 3-6-30, BU, SCR 11-10-53

4123 D 3-11-30, BU, wired for higher speed, ACC Superior and E. 140th, 12 injured on 9-21-43, operated in "Parade of Progress" 4-27-52, SCR 11-10-53

4124 G, D 3-20-30, BU, wired for higher speed, SCR 11-10-53

4125 G, D 3-12-30, SCR 10-29-53

4126 G, D 3-13-30, held for Shaker Rapid in 1953, rapid conversion work started but not taken, SCR 10-29-53
4127 G, D 3-18-30, SCR 10-29-53
4128 G, D 3-17-30, SCR 7-14-53
4129 G, D 3-20-30, SCR 10-29-53
4130 D 3-24-30, BU, SCR 11-10-53
4131 G, D 3-27-30, BU, SCR 3-30-54
4132 G, D 4-9-30, BU, wired for higher speed, four rubber bolsters and eight resilient wheels installed, REB the same as 4073 in 1939, SCR 3-30-54
4133 D 3-28-30, BU, SCR 1954
4134 G, D 4-2-30, OOS E. 55th 1950, SCR 1-17-52
4135 D 4-1-30, BU, ACC front end smashed in 1952, not repaired, SCR 3-30-54
4136 G, D 4-4-30, BU, had sign on side 1-24-54: "Lakewood's Last Streetcar Ride—Sponsored by Lakewood Chamber of Commerce," had flag-bunting, SCR 3-30-54

Modern bullet-eye and leather seats were the refinements to Cleveland's modern 4000-series streetcars after their 1939 rebuilding program. *(Cleveland Railway photo, Blaine Hays Collection)*

4137   G, D 4-11-30, BU, SCR 1954
4138   G, D 4-12-30, BU, had flag-bunting for last day on Madison 1-24-54, SCR 3-30-54
4139   D 4-15-30, BU, SCR 11-10-53
4140   D 4-17-30, BU, SCR 11-10-53
4141   D 4-18-30, BU, SCR 11-10-53
4142   G, D 4-22-30, BU, had flag-bunting and made final run on last day of streetcars on Madison 1-24-54, SCR 3-30-54
4143   G, D 4-23-30, BU, SCR 11-10-53
4144   D 4-26-30, BU, had flag-bunting for last day on Madison 1-24-54, sold to Norman Muller and moved to his residence in South Lorain, PTD green, had a whistle and pipe organ installed and fender from Southwestern interurban attached, had "Arlington Traction Co." on side, SCR 1962
4145   D 4-30-30, BU, had flag-bunting for last day on Madison 1-24-54, SCR 3-30-54
4146   D 5-3-30, BU, SCR 3-30-54
4147   D 5-7-30, BU, SCR 3-30-54
4148   G, D 5-14-30, wired for higher speed, SCR 11-10-53
4149   D 5-16-30, MOT rewound for higher speed on 12-11-31, fastest trolley in Cleveland at 50 mph, SCR 10-29-53

## PCC Cars in Cleveland
*by Bob Korach*

While Presidents' Conference Committee (PCC) cars, designed by the leadership of the national transit industry as the next generation of streetcars, were a major part of the planning process of most large transit systems in the late 1930s and all of the 1940s, Cleveland Railway and its successor Cleveland Transit System paid little attention to this greatly improved streetcar. In September 1934 the American Transit Association held its convention in Cleveland, and four new sample streetcars, including the first PCC, gave rides around town.

In November and December 1938, Pittsburgh PCC car 1095 (converted for its operation in Cleveland to two-man) roamed the Cleveland Railway lines in regular service. Vincent Sheldon operated this car on the Mayfield line westbound between Taylor and Lee at a speed never again to be equalled. Sheldon soon became a dispatcher, 1095 returned to Pittsburgh, and Cleveland Railway had done its duty.

Finally on October 1, 1946, fifty PCC cars entered service on Superior and St. Clair. Superior was supposed to be 100 percent PCC, but there were never enough cars to cover all peak service. St. Clair had PCC cars on a weekday basis and in the evenings and all day Sundays. Trailer trains gave all peak-hour service; two-man single cars gave Saturday service. The Nottingham branch of St. Clair did not receive PCC cars but base shuttle buses instead. The East 140th extension of Superior was replaced by a third branch of St. Clair using PCC cars and conventional trains.

In 1947 the twenty-five Louisville PCC cars arrived and went into service on St. Clair and Superior. On May 1, 1947, original PCC cars 4200–4219 went into base and some peak service on West 25th, State and Pearl branches. The Pullman PCC cars had backup controls needed to maneuver around Brooklyn Station. On August 1, 1947, the Pearl branch of West 25th Street was converted to bus, leaving State and later Broadview branches to some PCC operation mixed with rush-hour trailer trains. During summer 1947 Superior cars were again extended over East 140th Street to Euclid Beach on Sunday. On August 16, 1947, the Collinwood branch of St. Clair was changed to temporary bus, the Pullman PCC cars with their backup controls replacing shuttle buses to Nottingham. In 1948 the Collinwood branch of St. Clair went back to streetcars, and the East 140th branch was permanently changed to bus (May 5). During summer 1948 the Superior PCC cars went via Collinwood to Euclid Beach on Sunday only. They continued to do this in the summers of 1949 and 1950. On July 1, 1950, the Broadview branch of West 25th went to buses, and the PCC cars ran only on the surviving State Road branch. Also on July 1, 1950, 4213–4219 went into Clark base service and Lorain owl service between West 140th Street and Public Square, thus PCC cars on two more lines.

On February 17, 1951, Clark Avenue and Lorain owl service went back to standard cars (4000s) because of the inability of Denison Station to properly maintain the PCCs. On April 7, 1951, all St. Clair service east of East 129th Street (St. Clair Station) was converted to bus. On November 3, 1951, all streetcar service on St. Clair ended. On December 2, 1951, the displaced PCCs went into service on East 55th Street, using cars 4215–4232. West 25th Street had 4200–4214, and Superior retained 4233–4249 and Louisville 4250–4274, plus some old-type one-man cars on each line.

On April 27, 1952, a parade was held on Euclid Avenue commemorating the end of streetcar service on this most important line. PCC car 4215 appeared in the parade and was the only known run of a PCC on Euclid. On October 15, 1952, 4232 and 4251 were shipped to Toronto, Ontario, and by year-end 4250–4274 were also there.

By March 7, 1953, all of the PCC cars had left for Toronto, 4233 making a ceremonial trip on East 55th, ending streetcar service on that line and closing East 55th Station forever. The 4000-type cars continued on Superior until it went to trackless trolleys on March 20, 1953.

One little known fact about the original Pullman PCC cars was that they were to have gone in service on Union, Broadway, Fulton, and Clark. They finally made it to Clark, but that was all.

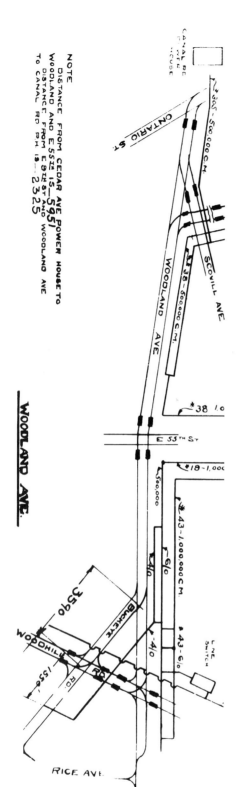

## Fleet Statistics: PCC Cars, 4200–4249

BLDR: Pullman-Standard (Worcester Plant); LEN: 46' 6"; seats: 53; HDL: ESSCO 32v; TRKS: Clark B-2; crew size: one; FEN: HB lifeguard; W: 39,800 lbs; MOT: 4 300v WH 1432J; CTL: WH XMA 452; CTS T:33. PCC cars on CTS were known as "Transitliners." All cars delivered to CTS by August 1946. All cars shipped to Toronto Transit Commission (TTC) in 1953 except where noted. The Toronto PCC rail-grinding train W-30 and W-31 PTD yellow with black striping, earning them the name "Banana Cars." All cars OOS in Toronto in 1982. This fleet purchased under first series equipment trust, dated 6-1-46.

| | |
|---|---|
| CTS 4200 | OOS 1-27-53, TTC 4625 |
| CTS 4201 | OOS 1-30-53, TTC 4626 |
| CTS 4202 | OOS 1-15-53, TTC 4627 |
| CTS 4203 | OOS 1-30-53, TTC 4628 |
| CTS 4204 | OOS 3-9-53, TTC 4629 |
| CTS 4205 | OOS 2-5-53, TTC 4630, returned to Cleveland RTA for Shaker service 9-6-78 |
| CTS 4206 | OOS 1-9-53, TTC 4631, CON to TTC rail grinder W-30, PTD yellow, still in TTC service |
| CTS 4207 | OOS 2-12-53, TTC 4632 |
| CTS 4208 | OOS 2-17-53, TTC 4633, preserved at Halton County Railway Museum, Ontario |
| CTS 4209 | OOS 3-17-53, TTC 4634 |
| CTS 4210 | OOS 12-18-52, TTC 4635, shipped 1952 |
| CTS 4211 | OOS 2-12-53, TTC 4636 |
| CTS 4212 | OOS 3-30-53, TTC 4637 |
| CTS 4213 | OOS 1-21-53, TTC 4638, had additional number PTD above destination sign |
| CTS 4214 | OOS 3-30-53, TTC 4639 |
| CTS 4215 | operated in "Parade of Progress" 4-27-52, only PCC ever on Euclid Ave., OOS 4-6-53, TTC 4640 |
| CTS 4216 | OOS 1-6-53, TTC 4641 |
| CTS 4217 | OOS 3-17-53, TTC 4642 |
| CTS 4218 | OOS 3-30-53, TTC 4643 |
| CTS 4219 | OOS 1-9-53, TTC 4644 |
| CTS 4220 | OOS 4-6-53, TTC 4645 |
| CTS 4221 | OOS 2-19-53, TTC 4646 |
| CTS 4222 | OOS 4-21-53, TTC 4647 |
| CTS 4223 | OOS 4-27-53, TTC 4648, returned to Cleveland RTA for Shaker service 8-15-78 |
| CTS 4224 | OOS 5-5-53, TTC 4649 |
| CTS 4225 | OOS 1-6-53, TTC 4650 |

Car 4213, PCC, Pullman-Standard, 1946, CTS type 33. *(Jim Spangler photo)*

| | |
|---|---|
| CTS 4226 | OOS 4-27-53, TTC 4651, returned to Cleveland RTA for Shaker service 9-8-78 |
| CTS 4227 | OOS 5-6-53, TTC 4652, returned to Cleveland RTA for Shaker service 8-8-78 |
| CTS 4228 | OOS 2-5-53, TTC 4653 |
| CTS 4229 | OOS 2-5-53, TTC 4654 |
| CTS 4230 | OOS 7-24-53, TTC 4655, returned to Cleveland RTA for Shaker service 9-19-78 |
| CTS 4231 | OOS 2-17-53, TTC 4656, returned to Cleveland RTA for Shaker service 10-25-78 |
| CTS 4232 | OOS 10-2-52, TTC 4657, shipped 10-15-52 |
| CTS 4233 | OOS 12-19-52, TTC 4658, shipped 1952 |
| CTS 4234 | OOS 2-6-53, TTC 4659 |
| CTS 4235 | OOS 12-11-52, TTC 4660, shipped 1952 |
| CTS 4236 | OOS 7-8-53, TTC 4661 |
| CTS 4237 | OOS 6-23-53, TTC 4662, returned to Cleveland RTA for Shaker service 11-3-78 |

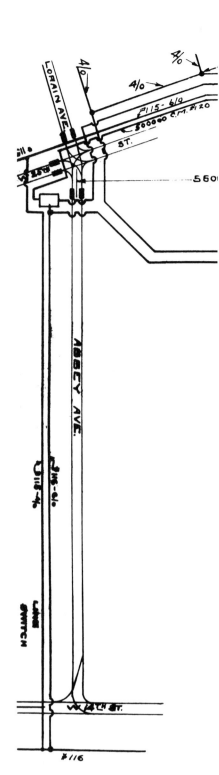

| | |
|---|---|
| CTS 4238 | OOS 12-18-52, TTC 4663, returned to Cleveland RTA for Shaker service 11-17-78 |
| CTS 4239 | OOS 6-5-53 |
| CTS 4240 | OOS 1-21-53, TTC 4665, returned to Cleveland RTA for Shaker service 12-19-78 |
| CTS 4241 | OOS 2-17-53, TTC 4666 |
| CTS 4242 | OOS 6-29-53, TTC 4667 |
| CTS 4243 | OOS 6-10-53, TTC 4668, CON to TTC rail grinder W-31, PTD yellow, still in TTC service |
| CTS 4244 | OOS 6-2-53, TTC 4669 |
| CTS 4245 | OOS 1-26-53, TTC 4670 |
| CTS 4246 | OOS 6-30-53, TTC 4671 |
| CTS 4247 | OOS 1-13-53, TTC 4672 |
| CTS 4248 | OOS 7-7-53, TTC 4673 |
| CTS 4249 | OOS 7-2-53, TTC 4674, extant on private property near Toronto, at riding stable |

### Fleet Statistics: PCC Cars, 4250–4274

BLDR: St. Louis Car Co.; LEN: 46' 5⅜"; seats: 58; HDL: ESSCO 32v; TRKS: Clark B-2; crew size: one; FEN: HB Lifeguard; W: 38,500 lbs; MOT: 4 WH 1432 HE 300v; CTL: WH XMA 2025; CTS T 34. These cars ACQ from Louisville Railway Co. (LRC) in September 1946 in a swap for buses and cash. Louisville got CTS GM coaches 3405–3424 in exchange for the 25 PCC cars. Cars 501–510 had already been delivered to Louisville, but the remainder were shipped directly to Cleveland. They were regauged, repainted, and REN in Cleveland in accordance with the PCC numbering pattern. Buses shipped to Louisville received the PCC numbers of the cars they replaced. PCC cars ACQ by Toronto for $17,500 each; all St. Louis cars were shipped in 1952. All cars OOS in Toronto in 1982.

| | |
|---|---|
| LRC 525 | CTS 4250, OOS 12-5-52, TTC 4675, had skirting cut |
| LRC 501 | CTS 4251, OOS 10-1-52, TTC 4676 |
| LRC 502 | CTS 4252, OOS 12-5-52, TTC 4677 |
| LRC 503 | CTS 4253, OOS 11-14-52, TTC 4678 |
| LRC 504 | CTS 4254, OOS 11-18-52, TTC 4679 |
| LRC 505 | CTS 4255, OOS 10-31-52, TTC 4680 |
| LRC 506 | CTS 4256, OOS 11-7-52, TTC 4681 |
| LRC 507 | CTS 4257, OOS 12-8-52, TTC 4682 |
| LRC 508 | CTS 4258, OOS 11-18-52, TTC 4683, had skirting cut |
| LRC 509 | CTS 4259, OOS 10-28-52, TTC 4684, preserved at Halton County Railway Museum, Ontario |

Car 4272, PCC, St. Louis, 1946, CTS type 34. *(Anthony F. Krisak photo, Richard Krisak Collection)*

| | | |
|---|---|---|
| LRC 510 | CTS 4260, OOS 11-11-52, TTC 4685 | |
| LRC 511 | CTS 4261, OOS 10-20-52, TTC 4686, had skirting cut | |
| LRC 512 | CTS 4262, OOS 11-19-52, TTC 4687 | |
| LRC 513 | CTS 4263, OOS 12-5-52, TTC 4688, had skirting cut | |
| LRC 514 | CTS 4264, OOS 12-5-52, TTC 4689, had skirting cut | |
| LRC 515 | CTS 4265, OOS 11-19-52, TTC 4690 | |
| LRC 516 | CTS 4266, OOS 11-21-52, TTC 4691 | |
| LRC 517 | CTS 4267, OOS 12-9-52, TTC 4692 | |
| LRC 518 | CTS 4268, OOS 10-4-52, TTC 4693, had skirting cut | |
| LRC 519 | CTS 4269, OOS 10-24-52, TTC 4694, had skirting cut | |
| LRC 520 | CTS 4270, OOS 12-9-52, TTC 4695 | |
| LRC 521 | CTS 4271, OOS 12-9-52, TTC 4696 | |
| LRC 522 | CTS 4272, OOS 10-28-52, TTC 4697 | |
| LRC 523 | CTS 4273, OOS 11-12-52, TTC 4698, had skirting cut | |
| LRC 524 | CTS 4274, OOS 11-12-52, TTC 4699, had skirting cut | |

LOANED PCC CARS

*Pittsburgh Railways Co. air PCC car 1095:* 1095 was loaned to CR in

1938 to introduce the public to modern electric transportation. Pressure was on CR to purchase the modern cars, but money was not there. Beginning operations on November 23, 1938, the car was marketed in Cleveland as the "Streetliner," and its operation resulted in upgrading several of the 4000-series cars to PCC quality. Standard gauge trucks were imported from Washington, D.C., for regauging the car while in Cleveland. 1095, the first Raymond Loewy green-gray streetcar, BLT by St. Louis, sat fifty-four people, had Clark B-2 trucks, four GE 1198 300-volt motors, and GE control.

*Model "A" PCC car:* One of four experimental cars brought to Cleveland in 1934 for operation at the American Transit Association convention. This car had been loaned to the Electric Railway Presidents' Conference Committee (ERPCC) by Brooklyn and Queens Transit Corp. as a test car. It was Twin Coach's second and final entry into the streetcar market. It was number 5200 and served the Presidents' Conference Committee in field testing; thus it became known as the Model "A" PCC car.

*Model "B" PCC car:* BLT for Brooklyn and Queens Transit by Pullman as car 5300, this test car embodied all of the research accomplished by the ERPCC during the three-and-a-half-year development period, the first fully constructed PCC car.

*Pullman-Chicago car:* Numbered 4001, car was dubbed the "Blue Goose" because of its blue and silver color. It was constructed using research accomplished by the ERPCC.

*Brill-Chicago car:* Numbered 7001, car was first modern streetcar built by seasoned BLDR, Brill, and forerunner to the competition of the PCC car, the Brilliner. On Monday morning, September 24, 1934, the models "A," "B," Brill, and Pullman cars operated on Cleveland streets for the pleasure of convention delegates. It was the only time in history these four historic cars operated together. National Broadcasting Company microphones entered the cars on Superior in front of Hotel Cleveland. The resulting broadcast was the first ever made from a streetcar.

SECTION 5

## Cleveland Streetcar Roster of Equipment
### (Passenger Trailer Cars)

TRAILER OPERATIONS

*by Bob Korach*

When the Cleveland Railway was formed on March 1, 1910, it received only two double-truck trailers, No. 7 and No. 8. The first purchase of trailers was in 1912, and trailer operation continued until June 16, 1951, when the last ones were retired from the East 55th Street line.

In 1914 wooden motor cars 300–313, 315–331, and 333 were converted to trailers. In 1915 334–341, 343–350, 352, 354, 356, 360, and 736 were also converted to trailers. This totaled fifty-three converted trailers. In 1915 because of the arrival of new 300–329; 300–313, and 315–329 were renumbered 670–683 and 685–699. They were retired or sold starting in 1917, with the last being gone by 1924. A further renumbering occurred in 1920 with the remaining converted trailers renumbered to 6 and 90–99.

In 1927 motors 1233–1235 were converted to trailers and used on the Shaker Rapid until July 20, 1930, when they were stored in Miles Station. They were sold to CIRR on May 15, 1940, and returned to Shaker for further storage. They finally returned to service during and after World War II.

The first retrenchment from trailer operation was on November 1, 1928, when the West 14th–Brooklyn line was rerouted over Clark Avenue to West 100th–Denison, with trailers being eliminated due to the one-car wye at that location. On June 16, 1929, Mayfield lost its few trailers.

With delivery of the last three 5000s (5025–5027) in 1929 and the last 4000s (4079–4149) in 1930, all pre-CR wooden cars, except dinkeys, were retired, including the original two trailers, by December 1931.

On September 29, 1946, Superior lost its trailers followed by Euclid on August 15, 1949 (articulateds stayed until April 27, 1952); West 25th on June 30, 1950; Detroit on July 31, 1950; St. Clair on April 6, 1951. The last trailers in use ran on East 55th until June 16, 1951. Only 2318, 2319, and 2365 escaped scrapping, having been sold to the Shaker Rapid.

Two efforts were made to have multiple-unit (MU) train operation of two cars during the streetcar era. The first MU train operated on East 55th Street from 1913 to 1923 with 775 and 777 making up the train. It would appear that no problems were found, but these two cars were among the first four-motored 700s to be retired.

The next try at MU-train operation (aside from the Shaker 1200 type) was at the end of World War II when Madison cars 4048 and 4049 were rebuilt for train operation. After tests at Harvard Shops, the pair returned to Madison as single cars. Car 4049 retained its front coupler and Eclipse fender for a short time. No reason was given for this trial or its success or lack thereof.

## Use of Trailers Expands
*by J. William Vigrass*

The Tayler grant, supervised by Peter Witt, got off to a rapid start with order and delivery of one hundred new center-door trailers in 1912, cars 2000–2099. These were the first of 300 of the same design, and

492 of the same floor plan. The new trailers were among the lightest cars per passenger ever built.

The 49-foot trailer with an empty weight of 26,340 pounds weighed but 538 pounds per foot. For the sixty-five seated passengers, the car weighed 405 pounds per passenger. For a full standing load totaling 125 passengers, weight was only 210 pounds each. Every element of the car was subject to weight analysis, one result being the use of 22-inch-diameter wheels.

One objective of train operation was increased throughput of downtown terminal loops and intersections. A train, needing to stop and start at each intersection or car stop, took just slightly longer than a single car to pass through an intersection; thus, total throughput was increased substantially.

Motor cars at the time had 4 x 40 hp traction motors, so for a train, average horsepower per axle fell to 20. Scheduled speed was slower but was an acceptable trade-off, particularly following introduction of Peter Witt's alternate stop plan, the so-called Skip-Stop plan. To speed trailer augmentation, fifty-two box motor cars of the early 300–360 series plus car 736 were converted to trailers in 1914–15 and used as late as 1922. Trailers 1998 and 1999 continued in use and were the original double-truck trailers.

Many lines had a very sharp peak for which trailers were ideal. In the off peak, single motor cars could accelerate rapidly up to their 25-mph running speed, which conformed to urban speed limits. Good performance would appeal to off-peak discretionary riders. In rush hours, with much higher travel demand, trailers were effective. Running time of a typical line was around forty minutes; a train might take four to ten minutes more. In either event, the car or train could not make a second round-trip in the same rush hour, so the slightly longer running time was not a serious penalty.

Savings were substantial. A train rather than two motor cars had a crew of three rather than four. Capital cost of a trailer was substantially less than for a motor car, perhaps half. In addition, investment in the 600-volt DC electric traction power system was less. Substation capacity was less. Feeder cables were less. Demand and energy charges by the electrical utility company were less. It was a true low-cost solution to meeting peak needs. The trade-off was lower speed. But at that time, before widespread ownership of automobiles, that was acceptable.

### Fleet Statistics: Railroad Roof Trailers, 2000–2299

BLDR: G. C. Kuhlman; LEN: 49' 3"; seats: 59; TRKS: Brill 67F W: 25,900 lbs; seating style: longitudinal and end; CR T T4. All cars SCR at E. 34th Station except where noted.

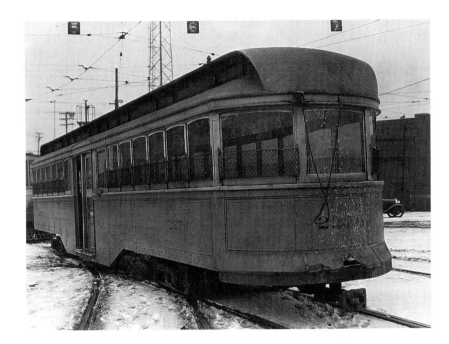

Trailer 2276, railroad-roof type,
G. C. Kuhlman, 1913, CR type T 4.
*(Cleveland Railway photo, Blaine Hays Collection)*

2000    D 9-14-12, SCR 1-13-36
2001    D 9-17-12, SCR 11-29-35
2002    D 9-14-12, CON to salt storage 9-4-34, SCR Harvard 4-6-40
2003    D 9-17-12, SCR 1-13-36
2004    D 9-14-12, SCR 1-13-36
2005    D 9-14-12, SCR 12-11-35
2006    D 9-14-12, SCR 11-29-35
2007    D 9-19-12, SCR 12-18-35
2008    D 9-19-12, SCR 12-10-35
2009    D 9-17-12, CON to salt storage 9-4-34, SCR 1-14-36
2010    D 9-20-12, SCR 12-3-35
2011    D 9-21-12, SCR 1-15-36
2012    D 9-21-12, SCR 12-20-35
2013    D 9-23-12, SCR 1-13-36
2014    D 9-23-12, SCR 12-9-35
2015    D 10-8-12, SCR 11-27-35
2016    D 9-24-12, SCR 11-19-35
2017    D 9-25-12, CON to salt storage 10-21-35, SCR Harvard 4-10-40
2018    D 9-25-12, SCR 1-18-36
2019    D 9-24-12, SCR 11-29-36
2020    D 9-25-12, SCR 12-19-35
2021    D 9-28-12, SCR 1-7-36
2022    D 9-28-12, SCR 12-13-35
2023    D 9-28-12, SCR 11-21-35
2024    D 9-28-12, SCR 1-14-36
2025    D 10-5-12, SCR 1-3-36

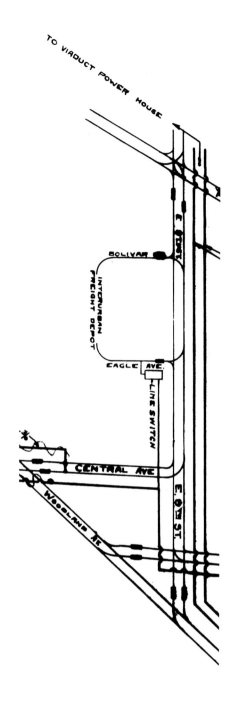

2026 D 9-28-12, SCR 12-12-35
2027 D 9-28-12, SCR 12-12-35
2028 D 9-28-12, SCR 11-27-35
2029 D 9-28-12, SCR 12-30-35
2030 D 10-25-12, CON to salt storage 10-21-35, stored at Cedar Station 1940, SCR Harvard 1941
2031 D 10-5-12, SCR 1-8-36
2032 D 10-5-12, SCR 1-14-36
2033 D 10-5-12, SCR 1-15-36
2034 D 10-5-12, SCR 1-6-36
2035 D 10-25-12, SCR 11-22-35
2036 D 10-5-12, SCR 1-14-36
2037 D 10-10-12, SCR 12-9-35
2038 D 10-10-12, SCR 11-19-35
2039 D 10-5-12, had imitation leather seats, SCR 1-6-36
2040 D 10-5-12, SCR 1-13-36
2041 D 10-18-12, SCR 1-9-36
2042 D 10-5-12, SCR 12-31-35
2043 D 10-5-12, CON to salt storage 10-21-35, SCR Harvard 4-17-40
2044 D 10-5-12, SCR 12-31-35
2045 D 10-5-12, SCR 1-4-36
2046 D 10-5-12, SCR 1-6-36
2047 D 10-8-12, SCR 1-16-36
2048 D 10-8-12, SCR 1-3-36
2049 D 10-5-12, CON to salt storage 10-21-35, SCR Harvard 4-15-40
2050 D 10-8-12, SCR 1-4-36
2051 D 10-5-12, SCR 12-21-35
2052 D 10-10-12, SCR 1-15-36
2053 D 10-5-12, SCR 1-10-36
2054 D 10-18-12, CON to salt storage 10-21-35, SCR Harvard 4-8-40
2055 D 10-10-12, CON to salt storage 10-21-35, SCR Harvard 4-12-40
2056 D 10-10-12, SCR 12-17-35
2057 D 10-18-12, SCR 11-29-35
2058 D 10-18-12, SCR 1-9-36
2059 D 10-18-12, SCR 1-11-36
2060 D 10-19-12, CON to salt storage 10-21-35, SCR Harvard 4-6-40
2061 D 10-25-12, CON to salt storage 10-21-35, SCR Harvard 4-15-40
2062 D 10-24-12, CON to salt storage 10-21-35, SCR Harvard 4-4-40
2063 D 10-24-12, SCR 12-20-35

| | |
|---|---|
| 2064 | D 10-19-12, SCR 1-9-36 |
| 2065 | D 10-19-12, SCR 12-13-35 |
| 2066 | D 10-19-12, SCR 1-9-36 |
| 2067 | D 10-19-12, SCR 1-10-36 |
| 2068 | D 10-19-12, SCR 1-8-36 |
| 2069 | D 10-25-12, SCR 1-9-36 |
| 2070 | D 10-29-12, SCR 1-6-36 |
| 2071 | D 10-31-12, SCR 1-11-36 |
| 2072 | D 11-4-12, SCR 11-29-35 |
| 2073 | D 11-6-12, CON to salt storage 10-21-35, SCR Harvard 4-1-40 |
| 2074 | D 11-7-12, SCR 12-21-35 |
| 2075 | D 11-17-12, SCR 1-7-36 |
| 2076 | D 11-17-12, SCR 12-21-35 |
| 2077 | D 12-3-12, SCR 12-14-35 |
| 2078 | D 12-2-12, SCR 1-15-36 |
| 2079 | D 12-5-12, SCR 1-8-36 |
| 2080 | D 11-29-12, SCR 12-20-35 |
| 2081 | D 11-29-12, SCR 1-10-36 |
| 2082 | D 12-7-12, SCR 1-9-36 |
| 2083 | D 12-12-12, SCR 1-8-36 |
| 2084 | D 12-12-12, SCR 1-10-36 |
| 2085 | D 12-16-12, SCR 1-8-36 |
| 2086 | D 12-16-12, SCR 1-8-36 |
| 2087 | D 12-16-12, SCR 1-10-36 |
| 2088 | D 12-16-12, SCR 12-16-35 |
| 2089 | D 12-23-12, SCR 12-20-35 |
| 2090 | D 12-28-12, CON to salt storage 10-21-35, SCR Harvard 4-11-40 |
| 2091 | D 1-3-13, SCR 12-20-35 |
| 2092 | D 1-3-13, SCR 1-6-36 |
| 2093 | D 1-8-13, SCR 12-18-35 |
| 2094 | D 1-8-13, SCR 12-11-35 |
| 2095 | D 1-8-13, SCR 12-21-35 |
| 2096 | D 1-14-13, SCR 12-20-35 |
| 2097 | D 1-16-13, SCR 12-18-35 |
| 2098 | D 1-21-13, SCR 12-16-35 |
| 2099 | D 1-31-13, SCR 12-20-35 |
| 2100 | D 3-4-13, SCR 1-6-36 |
| 2101 | D 3-4-13, SCR 12-17-35 |
| 2102 | D 3-4-13, SCR 12-2-35 |
| 2103 | D 3-5-13, CON to salt storage 10-21-35, SCR Harvard 4-1-40 |
| 2104 | D 3-5-13, SCR 1-6-36 |
| 2105 | D 3-5-13, CON to salt storage 10-21-35, SCR Harvard 4-17-40 |
| 2106 | D 3-5-13, SCR 12-30-35 |
| 2107 | D 3-6-13, SCR 12-12-35 |
| 2108 | D 3-6-13, SCR 1-15-36 |

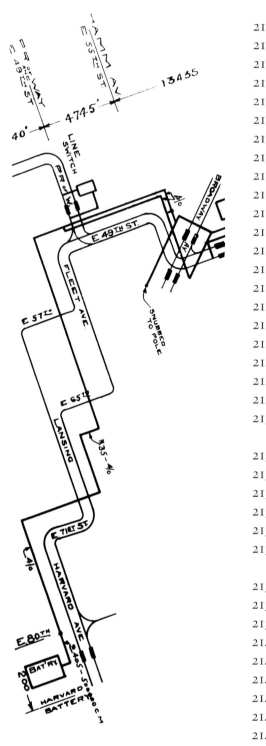

| | |
|---|---|
| 2109 | D 3-6-13, CON to salt storage 10-21-35, SCR Harvard 4-2-40 |
| 2110 | D 3-6-13, SCR 12-2-35 |
| 2111 | D 3-7-13, SCR 12-21-35 |
| 2112 | D 3-7-13, SCR 12-3-35 |
| 2113 | D 3-8-13, SCR 12-12-35 |
| 2114 | D 3-8-13, SCR 12-2-35 |
| 2115 | D 3-11-13, SCR 1-3-36 |
| 2116 | D 3-11-13, SCR 12-23-35 |
| 2117 | D 3-11-13, SCR 1-4-36 |
| 2118 | D 3-11-13, SCR 12-17-35 |
| 2119 | D 3-12-13, SCR 12-17-35 |
| 2120 | D 3-13-13, SCR 1-4-36 |
| 2121 | D 3-13-13, SCR 12-14-35 |
| 2122 | D 3-14-13, SCR 12-30-35 |
| 2123 | D 3-14-13, SCR 12-30-35 |
| 2124 | D 3-13-13, SCR 12-9-35 |
| 2125 | D 3-15-13, SCR 1-2-36 |
| 2126 | D 3-15-13, SCR 1-2-36 |
| 2127 | D 3-15-13, SCR 12-16-35 |
| 2128 | D 3-19-13, SCR 12-16-35 |
| 2129 | D 3-20-13, SCR 1-13-36 |
| 2130 | D 3-20-13, CON to salt storage 10-21-35, SCR Harvard 4-18-40 |
| 2131 | D 3-20-13, SCR 1-15-36 |
| 2132 | D 3-20-13, SCR 12-13-35 |
| 2133 | D 11-21-13, SCR 12-12-35 |
| 2134 | D 3-26-13, SCR 11-22-35 |
| 2135 | D 3-26-13, SCR 11-19-35 |
| 2136 | D 3-26-13, moved to E. 8th Street and Carnegie next to Service Building late in 1934, and never used, SCR 11-19-35 |
| 2137 | D 3-26-13, SCR 11-19-35 |
| 2138 | D 3-26-13, SCR 11-19-35 |
| 2139 | D 3-27-13, SCR 11-21-35 |
| 2140 | D 3-27-13, SCR 1-7-36 |
| 2141 | D 3-28-13, SCR 12-16-35 |
| 2142 | D 3-29-13, SCR 1-6-36 |
| 2143 | D 3-29-13, SCR 11-29-35 |
| 2144 | D 3-29-13, SCR 1-4-36 |
| 2145 | D 3-31-13, SCR 12-31-35 |
| 2146 | D 4-1-13, SCR 1-11-36 |
| 2147 | D 4-1-13, SCR 12-12-35 |
| 2148 | D 4-1-13, SCR 1-14-36 |
| 2149 | D 4-1-13, SCR 11-26-35 |
| 2150 | D 4-14-13, SCR 12-23-35 |
| 2151 | D 4-14-13, SCR 11-26-35 |
| 2152 | D 4-15-13, SCR 11-26-35 |

2153   D 4-15-13, SCR 12-2-35
2154   D 4-16-13, SCR 11-26-35
2155   D 4-16-13, SCR 1-7-36
2156   D 4-18-13, SCR 11-22-35
2157   D 4-18-13, SCR 11-27-35
2158   D 4-19-13, SCR 12-30-35
2159   D 4-22-13, SCR 12-23-35
2160   D 4-23-13, SCR 12-30-35
2161   D 4-25-13, SCR 11-27-35
2162   D 4-25-13, SCR 12-31-35
2163   D 4-25-13, SCR 12-30-35
2164   D 4-28-13, SCR 1-6-36
2165   D 4-29-13, SCR 11-27-35
2166   D 4-29-13, SCR 12-9-35
2167   D 5-2-13, SCR 1-14-36
2168   D 5-2-13, SCR 12-10-35
2169   D 5-2-13, SCR 12-10-35
2170   D 5-2-13, SCR 12-11-35
2171   D 5-3-13, SCR 1-4-36
2172   D 5-3-13, sent to Fairmount Station (Cedar Station) as a tool car in 1936 and parked in rear of barn; car remained there until 1949 when streetcar activity was transferred to Windermere; SCR in place 1949 as part of station cleanup, car OOS for 13 years.
2173   D 5-5-13, SCR 1-11-36
2174   D 5-5-13, SCR 12-3-35
2175   D 5-7-13, SCR 11-21-35
2176   D 5-7-13, SCR 11-21-35
2177   D 5-8-13, SCR 12-14-35
2178   D 5-9-13, SCR 12-13-35
2179   D 5-19-13, SCR 11-21-35
2180   D 5-10-13, SCR 12-14-35
2181   D 5-10-13, SCR 12-9-35
2182   D 5-13-13, SCR 12-14-35
2183   D 5-13-13, SCR 1-11-36
2184   D 5-15-13, SCR 12-13-35
2185   D 5-15-13, SCR 12-17-35
2186   D 5-16-13, SCR 12-3-35
2187   D 5-17-13, SCR 11-22-35
2188   D 5-17-13, SCR 12-19-35
2189   D 5-20-13, SCR 12-14-35
2190   D 5-23-13, SCR 12-14-35
2191   D 5-23-13, SCR 12-18-35
2192   D 5-23-13, SCR 12-3-35
2193   D 5-31-13, SCR 12-11-35
2194   D 5-29-13, SCR 12-19-35

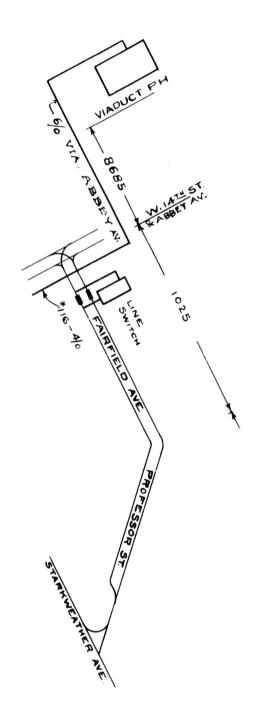

| | |
|---|---|
| 2195 | D 5-3-13, SCR 12-11-35 |
| 2196 | D 5-29-13, SCR 12-19-35 |
| 2197 | D 5-29-13, SCR 12-19-35 |
| 2198 | D 5-3-13, SCR 11-22-35 |
| 2199 | D 5-31-13, SCR 12-2-35 |
| 2200 | D 7-26-13, SCR 1-2-36 |
| 2201 | D 7-28-13, SCR 1-2-36 |
| 2202 | D 7-30-13, SCR 12-10-35 |
| 2203 | D 7-30-13, REN 2273 on 1-6-36 for salt storage use (was in better condition than the original 2273), REN 0640 in 1946 for Harvard Shops storage, SCR Harvard 3-29-54 |
| 2204 | D 8-2-13, SCR 1-7-36 |
| 2205 | D 8-2-13, had round forced air scoops on roof, SCR 12-3-35 |
| 2206 | D 8-2-13, SCR 1-7-36 |
| 2207 | D 8-2-13, REN 2257 on 1-13-36 because car was in better condition than 2257, CON to salt storage 4-10-40 and stored at Harvard, SCR Harvard 5-10-46 |
| 2208 | D 8-4-13, SCR 12-13-35 |
| 2209 | D 8-4-13, SCR 1-3-36 |
| 2210 | D 8-6-13, SCR 1-2-36 |
| 2211 | D 8-6-13, REN 2258 on 1-14-36 for salt storage use because car was in better condition than the original 2258, SCR Harvard 10-2-47 |
| 2212 | D 8-7-13, REN 2252 on 1-16-36 for salt storage use because car was in better condition than the original 2252, SCR Harvard 4-24-40 |
| 2213 | D 8-7-13, SCR 1-4-36 |
| 2214 | D 8-7-13, REN 2278 on 1-8-36 for salt storage use because car was in better condition than 2278, REN 0642 in 1946 and CON to section headquarters car, SCR Harvard 1-13-54 (also listed as SCR Harvard 3-25-54) |
| 2215 | D 8-12-13, SCR 12-19-35 |
| 2216 | D 8-12-13, SCR 1-3-36 |
| 2217 | D 8-13-13, SCR 11-22-35 |
| 2218 | D 8-13-13, SCR 1-2-36 |
| 2219 | D 8-13-13, SCR 1-3-36 |
| 2220 | D 8-13-13, REN 2260 on 1-8-36 because car was in better condition than 2260, CON to salt storage on 4-3-40, SCR Harvard spring 1947 |
| 2221 | D 8-13-13, SCR 12-2-35 |
| 2222 | D 8-15-13, SCR 12-11-35 |
| 2223 | D 8-15-13, SCR 1-14-36 |
| 2224 | D 8-15-13, SCR 1-14-36 |
| 2225 | D 8-16-13, REN 2290 on 1-6-36 because car was in better condition than 2290, CON to salt storage on 4-3-40, SCR Harvard spring 1947 |

2226  D 8-18-13, SCR 12-17-35
2227  D 8-18-13, SCR 12-23-35
2228  D 8-21-13, SCR 1-4-36
2229  D 8-21-13, SCR 12-21-35
2230  D 8-21-13, SCR 12-18-35
2231  D 8-21-13, SCR 12-23-35
2232  D 8-22-13, SCR 11-27-35
2233  D 8-22-13, SCR 12-16-35
2234  D 8-23-13, SCR 1-7-36
2235  D 8-23-13, SCR 11-26-35
2236  D 8-26-13, late in 1934 moved to Carnegie near E. 8th Street next to Service Building, never used, SCR at that location 12-4-35
2237  D 8-26-13, SCR 11-26-35
2238  D 8-27-13, SCR 12-18-35
2239  D 8-27-13, SCR 1-7-36
2240  D 8-28-13, SCR 12-23-35
2241  D 8-30-13, SCR 12-21-35
2242  D 8-30-13, SCR 12-21-35
2243  D 8-30-13, REN 2256 on 1-20-36 for salt storage because car was in better condition than 2256, SCR Harvard 4-26-40
2244  D 8-30-13, SCR 11-21-35
2245  D 8-30-13, SCR 12-10-35
2246  D 9-10-13, SCR 12-10-35
2247  D 9-10-13, SCR 12-21-35
2248  D 9-11-13, SCR 12-9-35
2249  D 9-12-13, SCR 12-23-35
2250  D 9-12-13, used for salt storage, OOS Denison 1947 (but was actually still in service by mistake until discovered in 1948), SCR Harvard 4-27-48
2251  D 9-12-13, SCR Harvard 10-1-47
2252  D 9-16-13, REN 2212 on 12-31-35 and SCR 1-3-36
2253  D 9-13-13, CON to salt storage 4-3-40, stored at Harvard 1940, SCR Harvard 4-25-46
2254  D 9-18-13, CON to salt storage 4-22-40, stored at Harvard 1940, SCR Harvard 4-2-46
2255  D 9-19-13, CON to salt storage 4-22-40, stored at Harvard 1940, SCR Harvard spring 1947
2256  D 9-20-13, REN 2243 on 1-16-36 and SCR 1-17-36
2257  D 9-25-13, REN 2207 on 1-10-36 and SCR 1-11-36
2258  D 10-22-13, REN 2211 on 1-7-36, SCR 1-15-36
2259  D 10-16-13, SCR Harvard 10-3-47
2260  D 10-24-13, REN 2220 on 1-9-36 and SCR on 1-10-36
2261  D 10-24-13, SCR Harvard 9-26-47
2262  D 10-22-13, SCR Harvard 4-22-40
2263  D 10-23-13, SCR Harvard 9-23-47

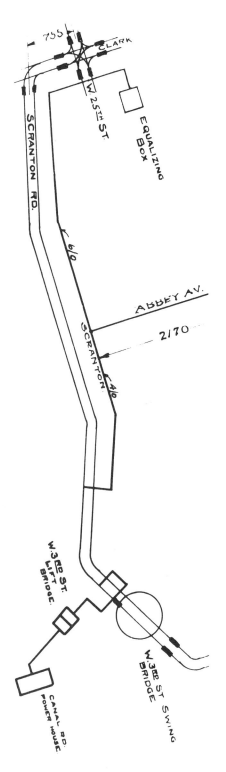

2264 D 10-23-13, SCR Harvard 9-25-47

2265 D 10-23-13, SCR Harvard 9-26-47

2266 D 10-24-13, CON to salt storage 4-8-40, stored at Harvard 1940 and SCR Harvard 4-25-47

2267 D 10-27-13, OOS Denison 1947, SCR Harvard 2-18-49

2268 D 10-24-13, CON to salt storage 4-15-40, stored at Denison 1940, SCR Harvard spring 1947

2269 D 10-24-13, SCR Harvard 4-24-40

2270 D 10-27-13, SCR Harvard 3-7-40

2271 D 10-27-13, SCR Harvard 9-26-47

2272 D 10-27-13, SCR Harvard 4-19-40

2273 D 10-27-13, REN 2203 on 12-31-35 and SCR 1-2-36

2274 D 10-29-13, SCR Harvard 4-8-40

2275 D 10-29-13, CON to salt storage 3-28-40, stored at Harvard 1940, SCR Harvard 4-8-46

2276 D 10-30-13, SCR Harvard 4-4-40

2277 D 10-30-13, CON to salt storage 4-8-40, stored at Harvard 1940, SCR Harvard spring 1947

2278 D 10-30-13, REN 2214 on 12-31-35 and SCR 1-3-36

2279 D 10-31-13, SCR Harvard 10-1-47

2280 D 10-31-13, REN 0641 in 1947 for section headquarters car, SCR Harvard 1-14-54

2281 D 10-31-13, CON to office at Harvard Yards for supervision of scrapping operations, SCR Harvard 9-23-47

2282 D 11-3-13, REN 0643 in 1947 for section headquarters car, SCR Harvard 12-5-52

2283 D 11-3-13, CON to salt storage 4-23-40, stored at Harvard 1940, SCR Harvard 9-30-47

2284 D 11-3-13, SCR Harvard 10-3-47

2285 D 11-5-13, REN 0644 in 1947 for section headquarters car, SCR Harvard 5-5-54, last trailer SCR at Harvard

2286 D 11-5-13, CON to salt storage 3-29-40, stored at Harvard 1940, SCR Harvard 4-18-41

2287 D 11-5-13, CON to salt storage 4-15-40, stored at Harvard 1940, SCR Harvard 1942

2288 D 11-17-13, SCR Harvard 9-30-47

2289 D 11-17-13, stored Superior 1940, SCR Harvard 3-8-41 (also reported SCR Harvard 3-19-41)

2290 D 11-17-13, REN 2225 on 12-30-35 and SCR 12-31-35

2291 D 11-19-13, had round forced air scoops on roof, SCR Harvard 9-23-47

2292 D 11-19-13, SCR Harvard 10-3-47

2293 D 11-21-13, SCR Harvard 10-1-47

2294 D 11-26-13, SCR Harvard 4-10-40

2295 D 11-26-13, OOS Denison 1947, SCR Harvard 2-18-49

2296 D 12-9-13, CON to salt storage 4-4-40, SCR Harvard 5-6-46

2297     D 12-1-13, CON to salt storage 4-15-40, SCR Harvard 4-15-41

2298     D 12-1-13, SCR Harvard 9-30-47

2299     D 12-1-13, CON to salt storage 4-4-40, SCR Harvard spring 1947

### Fleet Statistics: Arch Roof Trailers, 2300–2375

BLDR: 2300–2325 and 2351–2375, G. C. Kuhlman; 2326–2350, CR; LEN: 49'; seats: 59 (2300, 55); TRKS: Brill 67F; W: 25,900 lbs; CR T: 2300, T 5A; 2301–2375, T 5; 2376–2399, T 6. CR tried building some of these cars to determine if it could be done cheaper than Kuhlman, but Kuhlman was cheaper, as the following costs per car reflect: 2300–2325, $2,537.81; 2326–2350, $3,279.95; 2351–2375, $2,760.30. The car records list the following cars as having been CON to salt storage units—2337, 2339, 2341, 2342, 2353, 2354, 2357, 2366—but other sources dispute that assertion; the following list indicates both options.

Trailer 2300, arch-roof, G. C. Kuhlman, 1917, CR type T 5A. *(Cleveland Railway photo, Blaine Hays Collection)*

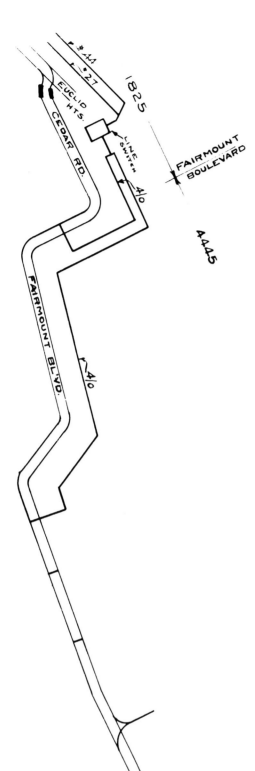

| | |
|---|---|
| 2300 | D 7-23-17, REB with 3 center doors in 1924, OOS 1948, SCR 6-4-48 |
| 2301 | G, D 7-30-17, OOS 1948, SCR 6-4-48 |
| 2302 | G, D 7-30-17, OOS St. Clair 1948, SCR 6-29-50 |
| 2303 | G, D 8-31-17, OOS St. Clair 1948, SCR 3-16-50 |
| 2304 | D 8-4-17, placed in dead storage at Denison 1940, reactivated during WWII, SCR 11-10-48 |
| 2305 | D 8-6-17, placed in dead storage at Denison 1941, reactived during WWII, SCR 3-25-49 |
| 2306 | D 8-6-17, OOS St. Clair 1948, SCR 11-25-49 |
| 2307 | D 8-17-17, placed in dead storage at Denison in 1940, reactived during WWII, SCR 4-22-49 |
| 2308 | D 8-18-17, OOS 1949, SCR 3-25-49 |
| 2309 | D 8-18-17, OOS 1949, SCR 1-28-49 |
| 2310 | D 9-13-17, OOS 1948, SCR 7-16-48 |
| 2311 | D 9-12-17, cost $3,380, OOS E. 55th 1948, SCR 10-16-50 |
| 2312 | G, D 9-18-17, OOS 1948, SCR 10-28-48 |
| 2313 | D 9-18-17, OOS 1949, SCR 3-21-49 |
| 2314 | D 9-14-17, placed in dead storage at Denison 1940, reactived during WWII, SCR 7-16-48 |
| 2315 | D 9-14-17, placed in dead storage at Denison 1940, reactivated during WWII, SCR 6-4-48 |
| 2316 | D 9-19-17, OOS 1948, SCR 7-16-48 |
| 2317 | D 9-20-17, destroyed by fire, SCR at E. 34th Station 1-15-36 |
| 2318 | G, D 9-20-17, sold to SHRT 10-28-46 |
| 2319 | D 9-21-17, sold to SHRT 10-28-46 |
| 2320 | G, D 9-29-17, OOS 1949, SCR 1-28-49 |
| 2321 | D 9-29-17, placed in dead storage at Denison 1941, reactivated during WWII, SCR 7-16-48 |
| 2322 | D 9-30-17, placed in dead storage at Denison 1941, reactivated during WWII, SCR 3-1-49 |
| 2323 | D 10-2-17, placed in dead storage at Denison in 1941, reactivated during WWII, SCR 11-29-49 |
| 2324 | D 10-2-17, destroyed by fire, SCR at E. 34th Station 1-13-36 |
| 2325 | D 10-9-17, OOS 1949, SCR 5-12-49 |
| 2326 | D 9-28-17, used as salt storage car at Madison Station for many years, stored at Denison Station in 1940, retrieved and renovated for passenger service during wartime in 1942, SCR 10-4-48 |
| 2327 | D 9-29-17, placed in dead storage at Denison 1941, reactivated during WWII, SCR 3-21-49 |
| 2328 | D 9-29-17, placed in dead storage at Denison 1940, reactivated during WWII, SCR 3-14-49 (inventory says SCR 3-21-49) |
| 2329 | D 10-4-17, placed in dead storage at Denison 1940, reactivated during WWII, SCR 6-4-48 |

2330   D 10-4-17, OOS St. Clair 1948, SCR 12-2-49
2331   G, D 10-5-17, OOS 1948, SCR 10-18-48
2332   D 10-8-17, OOS 1948, SCR 11-4-48
2333   G, D 10-8-17, OOS 1948, SCR 10-23-48
2334   G, D 10-9-17, OOS 1949, SCR 3-15-49
2335   D 10-11-17, placed in dead storage at Denison 1940, reactivated during WWII, moved to St. Clair 1948, SCR either 11-10-49 or 3-16-50
2336   D 10-11-17, placed in dead storage at Denison 1941, reactivated during WWII, moved to St. Clair 1948, SCR 11-25-49
2337   D 10-12-17, CON to salt storage at Harvard, stored Miles Station 1938, SCR 5-1-40
2338   G, D 10-17-17, OOS 1949, SCR 3-21-49
2339   D 10-18-17, CON to salt storage at Harvard, stored Miles Station 1938, SCR 5-6-40
2340   D 10-20-17, OOS 1948, SCR 10-28-48
2341   D 10-26-17, CON to salt storage at Harvard, stored Miles Station 1938, SCR 4-30-40
2342   D 10-26-17, CON to salt storage at Harvard, stored Miles Station 1938, SCR 4-29-40
2343   G, D 11-2-17, OOS E. 55th 1948, SCR 2-19-51
2344   G, D 11-3-17, OOS 1948, SCR 10-28-48
2345   G, D 11-7-17, OOS St. Clair 1948, SCR 11-25-49
2346   G, D 11-9-17, OOS St. Clair 1948, SCR 12-2-49
2347   G, D 11-14-17, OOS St. Clair 1948, SCR 12-14-49
2348   D 11-23-17, OOS 1948, SCR 10-22-48
2349   D 12-22-17, OOS 1948, SCR 11-10-48
2350   D 1-19-18, CON to salt storage at Harvard, moved to E. 55th 1948, SCR 8-9-51
2351   D 2-28-18, OOS 1948, SCR 10-20-48
2352   D 2-2-18, CON to salt storage at Harvard in 1948, SCR 9-30-52
2353   D 3-1-18, CON to salt storage at Harvard, stored Miles Station 1938, SCR 5-3-40
2354   D 3-3-18, CON to salt storage at Harvard, stored Miles Station, 1938, SCR 5-3-40
2355   G, D 3-28-18, OOS 1948, SCR 10-28-48
2356   G, D 3-5-18, OOS 1948, SCR 10-20-48
2357   D 3-26-18, CON to salt storage at Harvard, SCR 4-29-40
2358   G, D 3-23-18, OOS 1948, SCR 12-28-48
2359   G, D 3-4-18, OOS 1948, SCR 10-19-48
2360   G, D 3-4-18, OOS E. 55th 1948, SCR 7-18-50
2361   D 3-4-18, OOS 1948, SCR 10-23-48
2362   G, D 3-4-18, CON to salt storage at Harvard in 1948, SCR 11-9-51
2363   D 3-26-18, OOS 1948, SCR 11-4-48

Trailer 2365, arch roof, G. C. Kuhlman, 1917, CR type T 5. *(Blaine Hays photo)*

2364  G, D 3-2-18, OOS 1949, SCR 4-22-49

2365  G, D 3-2-18, sold to SHRT 10-26-46

2366  D 3-26-18, CON to salt storage at Harvard, stored Miles Station 1938, SCR 4-26-40

2367  D 3-23-18, made last revenue trailer run from E. 55th Station on CTS 6-16-51, SCR 9-30-52

2368  G, D 3-7-18, OOS 1949, SCR 5-12-49 (inventory says SCR 5-12-48)

2369  G, D 3-4-18, OOS 1949, SCR 3-21-49

2370  G, D 3-8-18, OOS 1949, SCR 4-5-49

2371  G, D 3-9-18, CON to salt storage at Harvard, moved to E. 55th Station in 1948, SCR 7-14-53

2372  G, D 3-8-18, OOS 1949, SCR 4-5-49

2373  G, D 3-8-18, OOS E. 55th 1949, SCR 6-15-51

2374  G, D 4-8-18, OOS 1949, SCR 3-21-49

2375  D 3-7-18, OOS 1949, SCR 4-22-49

### Fleet Statistics: Arch Roof Trailers, 2451–2500

BLDR: G. C. Kuhlman; LEN 49'; seats: 59; TRKS: Brill 67F; W: 25,900 lbs; CR T 7.

2451  G, D 7-29-20, OOS 1949, SCR 10-27-49 (inventory says SCR 1-11-50)

2452  G, D 7-29-20, OOS E. 55th 1949, SCR 7-5-51

2453  G, D 7-26-20, OOS 1951, SCR 8-21-51

2454  G, D 7-26-20, OOS 1951, SCR 6-25-51

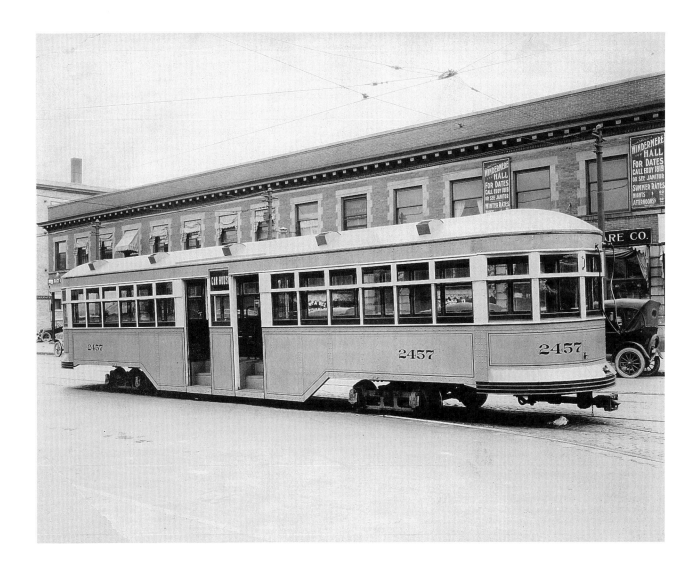

Trailer 2457, arch-roof type, G. C. Kuhlman, 1920, CR type T 7. *(Cleveland Railway photo, Blaine Hays Collection)*

2455    G, D 7-26-20, OOS E. 55th 1949, SCR spring 1951
2456    D 7-30-20, OOS 1950, SCR spring 1951
2457    G, D 7-30-20, OOS 1951, SCR 10-17-51
2458    G, D 7-30-20, OOS 1951, SCR 10-8-51
2459    D 8-6-20, OOS 1949, SCR 1-5-49
2460    D 8-6-20, OOS E. 55th 1948, SCR 10-15-51
2461    D 8-13-20, OOS E. 55th 1949, SCR 8-9-51
2462    G, D 8-13-20, OOS E. 55th 1948, SCR 1-23-52
2463    D 8-27-20, OOS E. 55th 1948, SCR 7-11-51
2464    G, D 8-16-20, OOS E. 55th 1949, SCR 5-15-51
2465    D 8-13-20, OOS 1949, SCR 5-19-49
2466    D 8-13-20, OOS 1951, SCR 11-21-51
2467    D 8-23-20, OOS 1951, SCR 1-23-52
2468    G, D 8-20-20, OOS E. 55th 1948, SCR 12-2-49
2469    G, D 8-19-20, OOS E. 55th 1949, SCR 10-29-51
2470    G, D 8-24-20, OOS E. 55th 1950, SCR 10-29-51

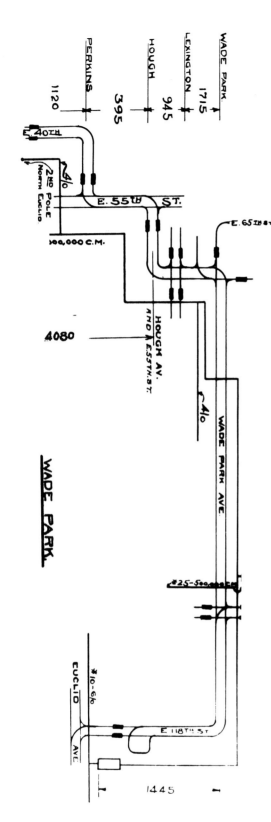

| | |
|---|---|
| 2471 | G, D 8-25-20, OOS E. 55th 1948, SCR 10-25-51 |
| 2472 | D 8-27-20, bad rear-end ACC, OOS 1951, SCR 6-29-51 |
| 2473 | D 8-31-20, OOS E. 55th 1949, SCR 8-13-51 |
| 2474 | G, D 8-31-20, OOS E. 55th 1949, SCR 7-5-51 |
| 2475 | G, D 9-2-20, OOS E. 55th 1949, SCR 8-10-51 |
| 2476 | D 9-8-20, OOS E. 55th 1949, SCR 6-28-51 |
| 2477 | D 9-8-20, OOS E. 55th 1950, SCR 6-20-51 |
| 2478 | D 9-9-20, OOS E. 55th 1949, SCR 8-8-51 |
| 2479 | G, D 9-9-20, OOS E. 55th 1949, SCR 12-5-51 |
| 2480 | G, D 9-2-20, OOS E. 55th 1949, SCR 7-9-51 |
| 2481 | D 9-11-20, OOS E. 55th 1950, SCR 1-3-52 |
| 2482 | G, D 9-11-20, OOS E. 55th 1949, SCR 8-8-51 |
| 2483 | G, D 9-14-20, OOS E. 55th 1948, SCR 8-14-51 |
| 2484 | G, D 9-15-20, OOS E. 55th 1949, SCR 11-12-51 |
| 2485 | D 9-15-20, OOS E. 55th 1950, SCR 1-17-52 |
| 2486 | D 9-17-20, OOS 1949, SCR 5-19-49 |
| 2487 | D 9-17-20, OOS E. 55th 1950, SCR 8-13-51 |
| 2488 | G, D 9-17-20, OOS E. 55th 1949, SCR 6-28-51 |
| 2489 | G, D 9-17-20, OOS E. 55th 1949, SCR 6-15-51 |
| 2490 | D 9-20-20, OOS E. 55th 1950, SCR 8-8-51 |
| 2491 | G, D 9-21-20, OOS E. 55th 1950, SCR 8-9-51 |
| 2492 | G, D 9-22-20, OOS E. 55th 1950, SCR 6-15-51 |
| 2493 | G, D 9-22-20, OOS E. 55th 1950, transferred without TRKS to Paramount Steel Co. across E. 49th Street (later Max Friedman) 8-7-51, SCR spring 1962 |
| 2494 | G, D 9-22-20, OOS 1950, SCR 5-15-51 |
| 2495 | D 9-25-20, OOS E. 55th 1950, SCR 3-18-51 |
| 2496 | D 9-29-20, OOS E. 55th 1949, SCR 4-26-51 |
| 2497 | G, D 9-28-20, OOS E. 55 1950, SCR 10-8-51 |
| 2498 | G, D 9-28-20, OOS 1949, SCR 5-31-49 |
| 2499 | G, D 9-30-20, CON to storage shed at Harvard in 1949, SCR 12-2-52 |
| 2500 | G, D 9-30-20, OOS E. 55th 1949, SCR 6-15-51 |

### Fleet Statistics: Center Entrance Trailers, 2376–2425

BLDR: Kuhlman; LEN: 49' 3"; seats: 59; TRKS: Brill 67F; W: 24,740 lbs; CR T T-6.

| | |
|---|---|
| 2376 | G, D 3-29-23, OOS E. 55th 1948, SCR 10-13-52 |
| 2377 | G, D 3-29-23, OOS E. 55th 1948, SCR 10-12-49 |
| 2378 | G, D 3-31-23, OOS E. 55th 1948, SCR 3-9-50 (inventory says SCR 10-25-49) |
| 2379 | G, D 3-31-23, OOS E. 55th 1948, SCR 10-12-49 |
| 2380 | G, D 3-31-23, OOS E. 55th 1948, SCR 10-12-49 |

2381  D 4-3-23, OOS 1949, SCR 3-9-50 (inventory says SCR 10-27-49)
2382  G, D 4-3-23, OOS 1951, 8-23-51
2383  G, D 4-4-23, OOS 1951, SCR 1951
2384  G, D 4-4-23, SCR 7-9-51
2385  G, D 4-5-23, OOS 1949, SCR 10-28-49
2386  G, D 4-5-23, OOS 1950, SCR 3-13-52
2387  G, D 4-6-23, OOS 1951, SCR 6-15-51
2388  G, D 4-6-23, OOS E. 55th 1949, SCR 7-21-50
2389  G, D 4-6-23, OOS E. 55th 1950, SCR 8-27-51
2390  G, D 4-9-23, OOS 1951, SCR 8-6-51
2391  D 4-9-23, OOS E. 55th 1950, SCR 1951
2392  G, D 4-10-23, OOS St. Clair 1950, SCR 8-15-51
2393  G, D 4-11-23, OOS 1951, SCR 10-9-51
2394  G, D 4-12-23, OOS 1949, SCR 5-19-49
2395  G, D 4-13-23, OOS 1951, SCR 10-8-51
2396  G, D 4-16-23, OOS 1951, SCR 9-12-51
2397  G, D 4-16-23, OOS St. Clair 1950, SCR 6-25-51
2398  G, D 4-19-23, OOS 1951, SCR 6-20-51
2399  G, D 4-19-23, OOS 1950, SCR 1951
2400  D, 4-19-23, OOS 1951, SCR 7-31-51
2401  G, D 4-27-23, OOS St. Clair 1950, SCR 6-20-51
2402  D 5-5-23, OOS 1951, SCR 6-15-51
2403  D 5-5-23, OOS St. Clair 1950, SCR 12-5-51
2404  G, D 5-8-23, OOS 1951, SCR 1-1951
2405  D 5-8-23, OOS 1951, SCR 10-8-51
2406  G, D 5-10-23, OOS St. Clair 1950, SCR 8-17-51
2407  G, D 5-12-23, OOS 1951, SCR 7-5-51
2408  G, D 5-16-23, OOS St. Clair 1950, SCR 7-31-51
2409  G, D 5-16-23, OOS 1949, SCR 12-14-49
2410  G, D 5-17-23, OOS 1949, SCR 12-14-49
2411  G, D 5-22-23, OOS St. Clair 1950, SCR 9-10-51
2412  G, D 5-19-23, OOS 1951, SCR 8-24-51
2413  G, D 6-14-23, OOS St. Clair 1950, SCR 6-25-51
2414  G, D 6-20-23, OOS 1951, SCR 1-1951
2415  D 6-21-23, OOS 1949 (in spite of this, car was still in use), SCR 7-31-51
2416  G, D 6-23-23, OOS 1951, SCR spring 1951
2417  G, D 6-23-23, OOS 1951, SCR 7-11-51
2418  G, D 6-26-23, OOS 1951, SCR 9-10-51
2419  G, D 6-27-23, OOS 1950, transferred without trucks to Paramount Steel Co. across E. 49th Street (later Max Friedman) 2-28-52, SCR spring 1969
2420  G, D 6-28-23, OOS 1951, SCR 8-28-51
2421  G, D 6-29-23, OOS 1949, SCR 2-16-49
2422  G, D 6-30-23, OOS 1951, SCR 1-1951

2423   G, D 6-30-23, OOS 1951, SCR 11-12-51
2424   G, D 6-29-23, OOS 1951, SCR 7-11-51
2425   G, D 6-30-23, OOS 1951, SCR 11-12-51

### Fleet Statistics: Arch Roof Trailers, 1982–1997, 1998, 1999

BLDR: CR; TRKS: B-67F; LEN: 50' 2"; W: 23,520 lbs; seats: 59, longitudinal and end; CR T 1.

1982   G, D 6-21-24, OOS 1951, SCR 11-21-51
1983   G, D 6-23-24, OOS 1951, SCR 10-23-51
1984   G, D 6-26-24, OOS 1951, SCR 10-17-51
1985   D 6-26-24, OOS E. 55th 1950, SCR 6-28-51
1986   D 7-1-24, LEN 51' 2", seated 61, OOS 1951, SCR 1-17-52
1987   D 7-3-24, LEN 51' 2", seated 61, OOS 1951, SCR 11-9-51

Trailer 1986, arch-roof type, Cleveland Railway, 1924, CR type T 1. *(Cleveland Railway photo, Blaine Hays Collection)*

| | |
|---|---|
| 1988 | D 7-8-24, LEN 51' 2", seated 61, OOS 1951, SCR 10-25-51 |
| 1989 | G, D 7-11-24, LEN 51' 2", seated 61, OOS 1950, SCR 3-27-51 |
| 1990 | G, D 7-16-24, LEN 51' 2", seated 61, OOS 1951, SCR 10-17-51 |
| 1991 | G, D 7-29-24, LEN 51' 2", seated 61, OOS 1951, SCR 10-29-51 |
| 1992 | G, D 3-3-24, OOS 1951, SCR 10-23-51 |
| 1993 | D 3-3-24, seated 61, OOS Harvard 1948, SCR 3-15-49 |
| 1994 | G, D 3-3-24, OOS 1951, SCR 11-21-51 |
| 1995 | G, D 3-3-24; leather seats (21 double, 7 single), 11 end Brill bus seats 1936, OOS E. 55th 1949, SCR 6-29-50 |
| 1996 | D 3-3-24, leather bucket seats installed in 1936, OOS 1951, SCR 11-1-51 |
| 1997 | D 3-3-24, Hale and Kilburn bus seats installed (21 double, 7 single, and 10 end) in 1936, OOS E. 55th 1949, SCR 6-29-50 |
| 1998 | BLT by Brill/Kuhlman in 1904 as number 7, REB in 1913 and REN in 1924 as 1998, LEN 38' 9", W 25,900, TRKS B67F (special), seats 16 cross, longitudinal and end, railroad roof, CR T 2, SCR 10-7-31 |
| 1999 | BLT by Kuhlman in 1904 as number 8, REB in 1913 and REN 1999 in 1924, LEN 37' 9", seats 52, 9 cross, longitudinal and end, railroad roof CR T 3, equipment removed and body sold 9-29-31 |

## SECTION 6
### *Service Cars*

*Units marked with asterisk were not numbered.*

| | |
|---|---|
| ★ | Windermere armature flat car: ST?, car never included in inventory, no date when built, SCR at E. 34th Station on 1-21-36 |
| ★ | Harvard Shops yard motor: BLT CER 1906, single Brill 21 TRK, box dinkey, SCR 10-7-42 |
| ★ | Mill motor flat car: one Dorner & Dutton TRK, BLT by CR 4-1-16, SCR 5-1-41 |
| ★ | rail bonder: BLT in 1908, removed from records 3-5-30 upon advice that car was discarded by Power Department some years prior to that, SCR 1921 |
| ★ | pay car: "Special" was passenger number 500, ST, BLT by Kuhlman in 1898 and had a Dupont TRK, LEN 32', car traveled to all stations with employee pay, 1934 report refers to car as "Special" and that was painted on side, SCR E. 34th Station 1-16-36 |
| 0 | mill room flat car, ST, had Brill #21 TRK, LEN 18', BLT by CER, SCR 5-1-41 |
| 00 | yard motor, ST, BLT in 1915, SCR 1940 |
| 00(2) | Pavement Plow, ST, SCR 5-1-41 |
| 01 | flat car, BLT 1895, SCR 1921 |

Pay car, G. C. Kuhlman, 1898, with roof-mounted headlight. *(Eugene Schmidt Collection, Northern Ohio Railway Museum)*

00, pavement plow in action. *(Cleveland Railway photo, Blaine Hays Collection)*

128 · CLEVELAND'S TRANSIT VEHICLES

| | |
|---|---|
| 1 | snow plow, ST?, BLT 1898 by Taunton Manufacturing Co. and had Taunton TRKS, two WH #49 MOT, two K-12 CTL, LEN 25', SCR at E. 34th station 1-18-36 |
| 01(2) | box freight, ST, DE, Dorner & Dutton TRK, BLT 1893 |
| 02 | box freight, ST, DE |
| 03 | box freight, ST, DE, originally number 3, later REN 0642, BLT by CER in 1903, LEN 24', Dorner & Dutton TRK, WH #93 MOT, K-10 CTL, car used for materials supply, car burned, SCR 3-20-41 (also reported SCR 3-22-41) |
| 04 | box freight, ST, DE, originally number 4, BLT by CER in 1902, LEN 23' 6", Dorner & Dutton TRK, WH #93 MOT, K-10 CTL, SCR 8-17-26 |
| 05 | box freight, ST, DE |
| 06 | box freight, ST, DE, originally number 6, BLT by CER 1902, LEN 24', had a Dorner & Dutton TRK, WH #93 MOT, K-10 CTL, car assigned to Harvard Shops for switching in later years, SCR 3-21-41 (also reported SCR 3-24-41) |
| 07 | box freight, ST, DE, originally number 7, BLT by CER in 1902, LEN 24', with Dorner & Dutton TRK, WH #93 MOT, K-10 CTL, SCR 8-17-26 |
| 08 | box freight, originally passenger car 358, REN 25, BLT by Brill in 1903, REB 1910 for work service, LEN 43', two Brill 27F TRKS, 4WH #101 MOT, K-6 CTL, SCR 7-26-29 |
| 09 | freight motor, REN 0644 (not to be confused with section headquarters trailer 2285, which received 0644 when 09 was SCR in 1946), LEN 24', BLT by CER 1905, Dorner & Dutton TRK, two WH #93 MOT, K-10 CTL, in later years assigned to Windermere Station as hose jumper equipment car, probably SCR late 1946 |
| 010 | box freight, ST, BLT 1905 by CER, LEN 24', Dorner & Dutton TRK, WH #93 MOT, K-10 CTL, REN 0645, in later years assigned to St. Clair Station as a hose jumper equipment car, SCR 3-1946 |
| 011 | box freight, ST, BLT 1906 by CER, LEN 24', Dorner & Dutton TRK, two WH #101 MOT, K-10 CTL, car was assigned in later years as a fender car for retrieving damaged "cow catcher" fenders, probably SCR late 1949 |
| 012 | dump freight, ST, BLT by CER in 1906, LEN 26' 6", Dorner & Dutton TRK, two WH #93 MOT, K-10 CTL, REN 0646 in later years and became a hose jumper equipment car at Harvard Shops, SCR 3-10-41 (also reported SCR 3-25-41) |
| 013 | flat motor sprinkler car, BLT by Miller & Noblock 1900, two Baldwin #427 TRKS, WH #101 MOT, two K-6-1 speed regulator CTL, MCB TRKS for moving railroad cars, FN 9 then REN 34 sometime before 1910, then REB by CR into a shop |

011, box freight, Cleveland Electric Railway, 1906. *(Cleveland Railway photo, Blaine Hays Collection)*

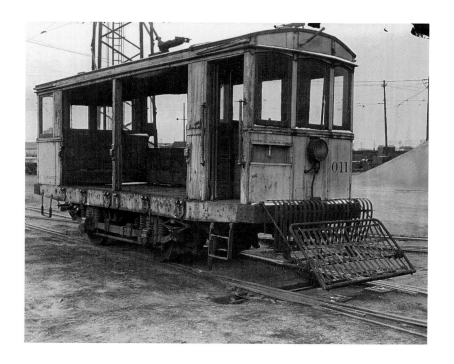

|  |  |
|---|---|
|  | locomotive in 1914 and REN 0760, became known to shop employees as "Jumbo," the big car weighed 85,100 lbs, SCR 6-10-54 |
| 014 | riveter car, BLT by CR March 1910, LEN 33', B-27F-1 TRKS, WH #49 MOT, K-12 CTL, REN possibly only in inventory to 0705, but appeared as 014 in CTS days, SCR 4-25-52 |
| 015 | flat motor (center cab), BLT by CR, D 1-20-13, LEN 35', 4WH #307F MOT, K-35-G speed regulator CTL, MCB standard TRKS, FN 3, REN 0761, performed interchange work with steam railroad, SCR 2-10-54 |
| 15 | fare box storage, former passenger 382, stationed at Superior starting in 1924 |
| 016 | dump freight, BLT 1913 by Orenstein Arthur Kopple, Standard TRKS, 4WH #305 MOT, HL CTL, car sold Differential Steel Car Co. 5-19-28 |
| 017 | dump freight, same as 016 except REN 0631, SCR 1951 |
| 018 | dump freight, same as 016 except sold to Differential Steel 5-1-29 |
| 019 | dump freight coal car, same as 016 except had Baldwin 4923 TRKS, 15B Master HL CTL 480F, later REN 0633, SCR 1951 |
| 020 | dump freight, BLT by Universal Dump Car Co. 8-2-14, LEN 41', Standard TRKS, 4WH #305 MOT, K35G CTL, sold to Differential Steel on 6-11-28 |
| 021 | dump freight, same as 020 except D 7-14-14, sold to Differential on 8-4-28 |

019, dump freight coal car, Orenstein Arthur Kopple, 1913. *(Cleveland Railway photo, Blaine Hays Collection)*

024(1)   welder, former passenger car 99, ST, parlor car built by CER 1893, LEN 32', 2WH12A MOT, K-10 CTL, became shelter house at St. Clair and Bliss Road, SCR 12-21-25

024(2)   welder?, FN 262, ST, CON in 1906 and SCR 1924

024(3)   welder, former passenger car 370, BLT 1899 by Brill, two B-27F TRKS, 4WH #101B-6 MOT, LEN 33', K-12 CTL, car REN 11 1910, REB to welder 1924, REN 024, later REN 0694, SCR 1947

024(4)   work motor, former passenger car 733, CON in 1926

034   pavement pounder, BLT 1912

037   flat car, LEN 20', Dorner & Dutton TRK, ST, BLT CER 1896, REN 0650, SCR E. 34th 1-18-36

038   flat car, LEN 20', Dorner & Dutton TRK, ST, BLT CER 1905, FN 31, SCR E. 34th 11-3-27

040   flat car, LEN 19', Dorner & Dutton TRK, ST, BLT CER 1902, FN 4, REN 0652, SCR E. 34th 1-20-36

044   flat car, LEN 19' 6", Dorner & Dutton TRK, ST, BLT CER 1902, FN 10, REN 0654, SCR E. 34th 1-20-36

047   flat car, LEN 20', Dorner & Dutton TRK, ST, BLT Kuhlman 1903, FN 8, REN 0655, SCR E. 34th 1-20-36

049   flat car, LEN 22', Dupont TRK, BLT CER 1896, FN 26, REN 0657, SCR E. 34th 1-18-36

| | |
|---|---|
| 050(1) | flat car, LEN 19', Dorner & Dutton TRK, ST, BLT CER 1896, FN6, REN 0658, SCR E. 34th 1-21-36 |
| 050(2) | flat car, accounted for as an "armature flat" and listed as never having been included in inventory, SCR E. 34th 1-21-36 |
| 051 | flat car, 19' 6", Dorner & Dutton TRK, ST, BLT Kuhlman 1903, REN 0659, SCR E. 34th 1-18-36 |
| 052 | flat car, LEN 19' 6", Dupont TRK, ST, BLT CER 1902, FN 7, REN 0660, SCR E. 34th 1-21-36 |
| 053 | flat car, LEN 19' 6", Dorner & Dutton TRK, BLT CER 1902, FN 2, SCR E. 34th 1-18-36 |
| 054 | steam roller flat, LEN 27', Union Engineering TRKS, BLT 1922 Union Engineering Co. as number 033, REN 0750, SCR 11-9-53 (also listed as SCR 10-9-53) |
| 055 | flat car, LEN 19', Dorner & Dutton TRK, ST, BLT CER 1896 as saw car number 15, REN 0770, SCR E. 34th 1-18-36 |
| 056 | flat car, LEN 22', Dupont TRK, BLT CER 1905, FN 28, REN 0661, SCR E. 34th 1-18-36 |
| 057 | flat car, LEN 19' 6", Dorner & Dutton TRK, ST, BLT 1902 CER, FN 3, REN 0662, SCR E. 34th 1-18-36 |
| 058 | flat car, LEN 19', Dorner & Dutton TRK, ST, BLT CER 1901, FN 12, REN 0663, SCR E. 34th 1-18-36 |
| 059 | flat car, LEN 22', Dupont TRK, ST, BLT 1905 CER, FN 30, REN 0664, SCR 7-10-47 |
| 060 | flat car, LEN 20', Dupont TRK, ST, BLT 1896 CER, FN 25, SCR 12-6-45 |
| 061 | flat car, LEN 19' 6", Dorner & Dutton TRK, ST, BLT 1901 CER, FN 13, REN 034, SCR E. 34th 1-20-36 |
| 063 | flat car, LEN 19' 6", Dorner & Dutton 25A TRK, ST, BLT W.B. Co. 1909, REN 0666, car was last of its type to be used in city streets in 1951, towed by Fender car 011, car SCR 7-10-51 |
| 064 | flat car, LEN 19', Dorner & Dutton TRK, ST, BLT 1909 W.B. Co., SCR 7-30-51 |
| 065 | flat car, LEN 19', Brill 21 TRK, ST, BLT W.B. Co. 1909, REN 0667, SCR 7-10-51 |
| 066 | flat car, LEN 20', Dupont TRK, BLT CER 1906, FN 035, REN 0668, SCR E. 34th 1-18-36 |
| 067 | flat car, LEN 19', Dorner & Dutton TRK, ST, BLT W.B. Co. Rec. 8-1909, SCR 1951 |
| 068 | flat car, LEN 19' 6", Dupont TRK, ST, BLT 1910 CR, REN 0669, SCR 7-10-51 |
| 069 | flat car, LEN 19', Dorner & Dutton TRK, ST, BLT W.B. Co., D 1909, REN 0670, SCR 7-30-51 |
| 070 | flat car, LEN 19', Dupont TRK, ST, BLT MT 1909, REN 0671, last surviving MT car, SCR E. 34th 1-21-36 |

072 flat car, LEN 19', Dorner & Dutton TRK, ST, BLT CR 1913, REN 0673, SCR 7-10-51

073 flat car, LEN 19', Dupont TRK, ST, BLT CR 1913, REN 0674, SCR E. 34th 1-18-36

074 flat car, LEN 19', Dorner & Dutton TRK, ST, BLT CR 1913, inventory record states that CR's Elmer Haber claimed this car was REN 061, probably after the original 061 was SCR at E. 34th on 1-20-36, car REN 0675 and probably lasted through war, SCR 1947?

076 flat car, LEN 19', Dorner & Dutton TRK, ST, BLT CR 1913, SCR E. 34th 1-18-36

077 dump trailer, LEN 34' 6", Brill 27F TRKS, BLT Orenstein Arthur Kopple Co. 1914, sold to Differential Steel Car Co. 5-1-29

078 dump trailer, LEN 34' 6", Standard TRKS, same as 077

079 dump trailer, same as 078

080 dump trailer, same as 078 except BLT 1913, sold to Differential 5-19-28

081 dump trailer, same as 078 except sold to Differential 5-1-29

082 dump trailer, same as 078

083 dump trailer, same as 078 except sold to Differential 5-19-28

084 dump trailer, same as 078 except had Brill 27F TRKS, sold to Differential 5-1-29

085 dump trailer, LEN 41', Standard TRKS, BLT Universal Car Co., D 8-7-14, sold to Differential 6-11-28

086 dump trailer, same as 078

087 dump trailer, same as 078 except sold to Differential 8-4-28

088 dump trailer, same as 078

096 flat car, LEN 19', Dorner & Dutton TRK, ST, BLT CER 1902, FN 1, REN 026, later REN 0677, SCR E. 34th 1-18-36

098 flat car, LEN 18', Dupont TRK, ST, BLT CER 1896, FN 16, REN 028, SCR 1951

099 flat car, LEN 18' 6", Dorner & Dutton TRK, ST, BLT CER 1896, FN 22, REN 029, later REN 0678, SCR E. 34th 1-21-36

0100 flat car, LEN 22', Dupont TRK, ST, BLT by CER 1905, FN 29, REN 030 (possibly also numbered 0222 for a time), and later REN 0679, SCR 7-24-53

0101(1) sweeper, ST, LEN 27', BLT Brill 1899, two K-10 CTL, one R-17 CTL, FN 3, REN XXVI, SCR 11-12-25

0101(2) sweeper, ST, LEN 28' 3", BLT McGuire, D 10-30-25, had 3WH #307F MOT, one K-12 CTL and two K-10 CTL, assigned to St. Clair Station, SCR 4-15-53

0102 sweeper, ST, LEN 27', BLT McGuire in 1904, FN 12, 3WH #49 MOT, two K-10 CTL and one R-17 CTL, assigned to Cedar Car House

0103 sweeper, ST, LEN 27', BLT McGuire 1905, FN 17, 3WH #49 MOT, 3 K-10 CTL, assigned to Rocky River Station, SCR 11-10-52

0104 sweeper, ST, LEN 24', BLT McGuire 1905, FN 18, two K-12 CTL and one K-10 CTL, 3WH #49 MOT, assigned to Woodhill Station, SCR 11-14-51

104 fare box storage, former passenger 104, assigned to Denison Station in 1923

0105 sweeper, ST, LEN 27', BLT McGuire 1903, FN 11, three K-10 CTL, 3WH #49 MOT, assigned to Woodhill Station, SCR 4-25-52 (inventory also says SCR 10-5-51)

0106 sweeper, ST, LEN 27' 6", BLT McGuire 1903, FN 2, three K-10 CTL, 3WH #49 MOT, assigned to Windermere Station, SCR 3-30-53

0107 sweeper, ST, LEN 28' 2", BLT McGuire 1903, FN 4, three K-10 CTL, 3WH #49 MOT, assigned to Brooklyn Station, SCR spring 1948

0108 sweeper, ST, LEN 28' 2", BLT McGuire 1903, FN 5, three K-10 CTL, 3WH #49 MOT, assigned to Denison Station, SCR 12-6-51

0109(1) sweeper, ST, LEN 26' 6", BLT Brill 1904, FN 16, three K-10 CTL, 3WH #49 MOT, car SCR 11-12-25

0109(2) sweeper, ST, LEN 28' 3", BLT McGuire, D 10-30-25, two K-12 CTL, 3WH #307F MOT, assigned to Windermere Station, SCR fall 1951

0110 sweeper, ST, LEN 25', BLT McGuire, D 10-27-14, three K-10 CTL, 3WH #49 MOT, assigned to Denison Station, SCR 12-10-53 (certain parts sold)

0111 sweeper, ST, LEN 28' 3", BLT McGuire, D 11-10-25, three K-12 CTL, 3WH #307F MOT, assigned to Denison Station, SCR 12-10-53 (certain parts sold)

0112 sweeper, ST, LEN 25', BLT McGuire 1900, FN 1, three K-10 CTL, 3WH #101 MOT, assigned to Cedar Car House, SCR 10-27-49

0113 sweeper, ST, LEN 25', BLT McGuire 1900, FN 9, three K-10 CTL, 3WH #49 MOT, assigned to Superior Station, SCR 10-10-51

0114 sweeper, ST, LEN 30', BLT McGuire 1902, FN 7, later XXIII, two K-12 CTL one K-10 CTL, 3WH #49 MOT, assigned to Woodhill Station, SCR fall 1951

0115 sweeper, ST, LEN 28' 3", BLT McGuire, D 11-15-15, three K-10 CTL, 3WH #49 MOT, assigned to E. 55th Station, SCR 12-10-53 (certain parts sold)

0116(1) sweeper, ST, BLT Brill, SCR 1911

0116(2) sweeper, ST, LEN 28', BLT Kuhlman, D 11-1911, FN 0126, REN in 1913 upon arrival of McGuire Sweeper 0126, three

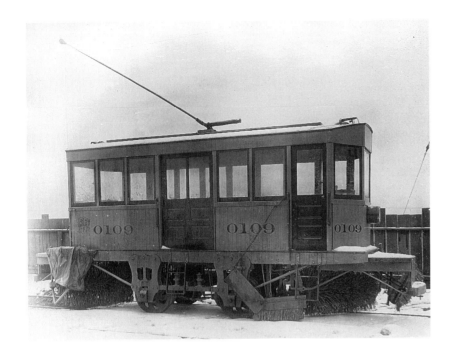

0109 (1), FN 16, snow sweeper, Brill, 1904. *(Cleveland Railway photo, Blaine Hays Collection)*

0113, FN 9, snow sweeper, McGuire, 1900. *(Cleveland Railway photo, Blaine Hays Collection)*

|  | K-10 CTL, 3WH #49 MOT, assigned to Harvard Shops, SCR 11-9-51 |
|---|---|
| 0117 | sweeper, ST, LEN 28' 3", BLT McGuire, D 11-12-12, three K-10 CTL, 3WH #101 MOT, assigned to Superior Station, SCR 7-24-53 |
| 0118 | sweeper, ST, LEN 28' 3", BLT McGuire, D 11-14-12, three K-10 CTL, 3WH #49 MOT, assigned to Woodhill Station, SCR 3-1-49 |
| 0119 | sweeper, ST, LEN 28' 6", BLT McGuire 1908, three K-10 CTL, 3WH #101 MOT, assigned to Rocky River Station, SCR 12-10-53 (certain parts sold) |
| 0120 | sweeper, ST, LEN 28' 6", BLT McGuire 1908, three K-10 CTL, 3WH #49 MOT, assigned to Superior Station, SCR 5-29-53 |
| 0121 | sweeper, ST, LEN 28' 6", BLT McGuire 12-1908, three K-10 CTL, 3WH #101 MOT, assigned to Woodhill Station, prepared for SCR 1952 and had numbers painted out, then kept in service at Harvard Shops to keep tracks clear, had numbers hand painted back on, SCR 1-13-54 |
| 0122 | sweeper, ST, LEN 28', BLT McGuire, D 12-30-13, three K-10 CTL, 3WH #49 MOT, at Madison Station, SCR 11-10-52 |
| 0123 | sweeper, ST, BLT McGuire, D 12-4-10, 1 K-12 and 2 K-10 CTL, 3EH #49 MOT, at Woodhill Station, SCR 7-24-53 |
| 0124 | sweeper, ST, LEN 28' 3", BLT McGuire, D 12-4-10, three K-10 CTL, 3WH #101 MOT, assigned to St. Clair Station, SCR spring 1953 |
| 0125 | sweeper, ST, LEN 28', BLT Kuhlman 11-1911, three K-10 CTL, 3WH #101 MOT, assigned to Woodhill Station, SCR 3-30-53 |
| 0126 | sweeper, ST, LEN 28', BLT McGuire, D 11-20-13, three K-10 CTL, 3WH #307 MOT, at E. 55th Station, SCR 10-30-52 |
| 0127 | sweeper, ST, LEN 25', BLT McGuire, D 10-27-14, three K-10 CTL, 3WH #49 MOT, assigned to St. Clair Station, SCR late in 1952 |
| 0128 | sweeper, ST, LEN 28' 3", BLT McGuire, D 11-15-15, three K-10 CTL, 3WH #49 MOT, assigned to Woodhill Station, SCR 11-15-52, thirty-seven years to the day after delivery! |
| 0129 | sweeper, ST, BLT McGuire, D 12-10-16, three K-10 CTL, 3WH #49 MOT, at Windermere Station, SCR 7-14-53 |
| 0130 | sweeper, ST, LEN 28' 3", BLT McGuire, D 12-8-17, three K-10 CTL, 3WH #49 MOT, at Superior Station, SCR 5-29-53 |
| 0131 | sweeper, ST, LEN 28' 3", BLT McGuire, D 12-8-17, three K-10 CTL, 3WH #49 MOT, assigned to Brooklyn Station, SCR 12-10-53 (certain parts sold) |
| 0132 | sweeper, ST, LEN 28' 3", BLT McGuire 11-1919, three K-10 CTL, 3WH #49 MOT, assigned to E. 55th Station, SCR 12-10-53 (certain parts sold) |

0133    sweeper, ST, BLT McGuire 11-1919, three K-10 CTL, 3WH #49 MOT, assigned to E. 55th Station, SCR 12-6-51

0134    sweeper, ST, LEN 28' 5", BLT McGuire, D 12-14-22, three K-10 CTL, 3WH #101 MOT, assigned to Denison Station, SCR 4-15-53

0135    sweeper, ST, LEN 28' 5", BLT McGuire, D 12-14-22, three K-10 CTL, 3WH #49 MOT, assigned to Denison Station, SCR 12-10-53 (certain parts sold)

0136(1) wrecker, ST, LEN 29', BLT Jones & Son 1890 as CER passenger number 248, Dorner & Dutton TRK, K-10 CTL, 2WH #49 MOT, assigned to Rocky River Station, SCR at E. 34th 1-17-36

0137(1) wrecker, ST, LEN 30', BLT Brill 1890 as passenger 310, Brill 21 TRK, 2WH #49 MOT, K-10 CTL, assigned to Lorain Station, SCR at E. 34th 1-16-36

0138    wrecker, ST, LEN 30', BLT Brill 1890 as passenger 223, Brill 21 TRK, 2WH #49 MOT, K-10 CTL, assigned to Superior Station, SCR at E. 34th 1-16-36

0139    wrecker, ST, LEN 30', BLT Brill 1890 as passenger 127, Brill 21 TRK, 2WH #49 MOT, K-10 CTL, assigned to Madison Station, SCR at E. 34th 1-16-36

0140    wrecker, ST, LEN 30', BLT Brill 1890 as passenger 165, Brill 21 TRK, 2WH #49 MOT, K-10 CTL, assigned to Miles Car House. Originally Woodland & Lorain car, transferred in 1916 from Miles to Harvard for use as training car, remained inside there for years and when finally "discovered" was an anachronism. On 10-1-46 brought out of hiding for parade heralding start of PCC operations on Superior; 1948 car was dusted off again for civic events and appeared on Public Square and in neighborhoods promoting transit usage; final great moment masquerading as horse car leading 4-27-52 "Parade of Progress" commemorating demise of streetcars on Euclid Avenue; returned to Harvard and sat in SCR line but was saved by Henry Ford Institute which ACQ it for $1.00 2-23-54; now on display with "Woodland & Lorain" on the clerestory, at Henry Ford Museum and Greenfield Village, Dearborn, MI.

0141    wrecker, ST, LEN 30', BLT Brill 1890 as passenger 169, Brill 21 TRK, 2WH #49 MOT, K-10 CTL, assigned to Cedar Car House, SCR at E. 34th 1-21-36

0142    wrecker, ST, LEN 30', BLT Brill 1890 as passenger 312, Brill 21 TRK, 2WH #49 MOT, K-10 CTL, REN 126 and became Holmden Car House wrecker, then REN 0131 and became a spare wrecker, SCR at E. 34th 1-16-36

0143    wrecker, ST, LEN 30', BLT Brill 1890 as passenger 304, later REN 130 and became work car 0134, Brill 21 TRK, 2WH

0147, wrecker, J. G. Brill, 1894 as passenger 362. *(Eugene Schmidt Collection, Northern Ohio Railway Museum)*

| | |
|---|---|
| 0147 | wrecker, ST, LEN 30', BLT Brill 1894 as passenger 362, later REN 127 and became work car 0132, Brill 21 TRK, 2WH #49 MOT, K-10 CTL, photos show car at Denison Station, SCR E. 34th 1-16-36 |
| 0148 | wrecker, ST, LEN 30', BLT Brill 1890 as passenger 128, later REN 154 and became wrecker 0133, assigned to St. Clair Car House, SCR 1-14-25 |
| 0149 | wrecker, ST, FN passenger 500, possibly same as Pay Car, nothing else known |
| 0149(2) | wrecker, LEN 41' 3", BLT Brill 1899 as passenger 734, REB in 1910 as 0149, Brill 27F TRKS, WH #101 MOT, K-10 CTL, assigned to Cedar Car House and stored there until 1944, moved to Harvard and SCR 4-24-45 |
| 0150(1) | power supply car, BLT Brill 1896 as passenger 268, Brill 21 TRK, WH12 MOT, K-10 CTL, CON in 1913 and SCR 1922 |
| 0150(2) | power supply car, LEN 41' 8", BLT Niles 1903 as passenger 108, in work service 1922, Brill 27F TRKS, WH #101 MOT, K-12 CTL, SCR 6-27-51 |
| 0151 | derrick, LEN 33', BLT CER 1906, B27F TRKS, 1WH 15 hp and 4WH340 MOT, K-6 CTL, SCR 11-21-25 |
| 0151(2) | rail bond test car, LEN 42' 8", BLT Niles 1903 as passenger 197, REB on 8-11-28 into 0151, B27F TRKS, 4WH #101 MOT, K-12 CTL, SCR 12-6-51 |
| 0152 | meter car, LEN 41' 3", BLT Brill 1899 as passenger 768, REB 1910 to meter car, SCR in late 1946 |

(The top of the first entry, partially shown above the caption area:) #49 MOT, K-10 CTL, assigned to Woodhill Station, SCR at E. 34th 1-18-36

0150 (2), power supply car, Niles, 1903, as passenger 108. (*Jim Spangler Collection*)

0155    sand car, ST, former passenger CON to sand car

0160    derrick, LEN 19', ST, BLT 1897 CER as #1, then REN 7, then REN 0152 then REN 0838, Dorner & Dutton TRK, 2WH #49 and 1 GE 12 hp MOT, WH K-10, R-14 CTL, SCR 11-26-52

0167    line car, ST, LEN 24' 4", BLT CER 1906, Dorner & Dutton TRK, 2WH #49 MOT, K-10 CTL, SCR 10-18-45

0168    line car, ST, LEN 23', BLT CER 1902 as #1, had Dupont TRK, 2WH #49 MOT, K-10 CTL, SCR E. 34th 1-16-36

0169    line car, LEN 38' 6", BLT Niles 1903 as passenger 112, Brill 27F TRKS, WH #101 MOT, K-12 CTL, REB 4-3-08 into mail car 0204 for Cleveland Circuit Railway Post Office and in 1919 under shop order #183 REB to line car 0169, SCR 11-26-52

0197    supply car, LEN 30', BLT CER as #XXII 1901, Brill 27G TRKS, WH #101 MOT, K35G CTL, made rounds of all car barns carrying general supplies and frequently pulled one of the flat cars, SCR 9-20-51

0207    flusher, ST, LEN 21', iron tank car, BLT Miller & Noblock as #15 1901, REN 077 and later 0207, Dupont TRK, 2WH #49 MOT, 1WH #60 type S.K. 15 hp MOT, K-10 CTL, on 8-9-29 Highway Oil Co. removed tank from the car at E. 55th Station, car SCR at E. 34th 1-17-36

0208    flusher, LEN 18', ST, iron tank car, BLT McGuire 1902 as #2, REN 087 and later 0208, Dorner & Dutton TRKS, 2WH #49 and 1WH #60L type S.K. 15 hp MOT, K-10 CTL, tank removed prior to SCR, SCR at E. 34th 1-17-36

0209  flusher, ST, LEN 21', iron tank car BLT CER 1903 as #11, REN 079, Dupont TRK, 2WH #49 and 1WH #60 type S.K. 15 hp MOT, K-10 CTL, on 8-9-29 tank removed by Highway Oil Co. at E. 55th Station, SCR at E. 34th 1-18-36

0210  flusher, ST, LEN 18', iron tank car BLT McGuire 1903 #3, REN 069, Dorner & Dutton 25A TRKS, 2WH #49 and 1WH #60 type S.K. 15 hp MOT, K-10 CTL, tank removed prior to SCR, SCR at E. 34th 1-17-36

0211  flusher, ST, LEN 21', iron tank car, BLT CER 1903 as #10, REN 081, Dupont TRKS, 2WH #49 and 1WH #60 type S.K. 15 hp MOT, K-10 CTL, 8-9-29 Highway Oil Co. removed tank at E. 55th Station, SCR at E. 34th 1-17-36

0212  flusher, ST, LEN 21', iron tank car, BLT CER 1904 as #6, REN 082, Dorner & Dutton TRKS, 2WH #49 and 1WH #60 S.K. 15 hp MOT, K-10 CTL, 8-9-29 Highway Oil Co. removed tank at E. 55th Station, SCR at E. 34th 1-17-36

0213  flusher, ST, FN 083, BLT Miller & Noblock 1903, SCR 1930

0214  flusher, ST, FN 084, BLT CER 1903, SCR 1929

0215  flusher, ST, iron tank car, BLT Brill 1896 as #1, REN 085, Brill 21E TRKS, 2WH #49 and 1WH 60L type S.K. 15 hp MOT, K-10 CTL, SCR at E. 34th 1-17-36

0216  flusher, ST, FN 086, BLT Miller & Noblock 1899, SCR 1926

0217  flusher, ST, LEN 21', iron tank car, BLT CER 1903 as #12, REN 078, Dupont TRKS, 2WH #49 and 1WH #60L type S.K. 15 hp MOT, two K-10 CTL, tank removed before car SCR at E. 34th 1-17-36

0218  flusher, ST, LEN 21', iron tank car, BLT CER 1903 as #9, REN 088 and later REN 0773, Dupont TRKS, 2WH #49 and 1WH #60 type S.K. 15 hp MOT, K-10 CTL, tank removed before SCR at E. 34th 1-17-36

0219  flusher, ST, LEN 21', iron tank car, BLT McGuire 1903 as #7, REN 089, Dupont TRK, 2WH #49 MOT, K-10 CTL, SCR 3-21-41 (also listed as SCR 3-15-41)

0220  flusher, ST, LEN 21', iron tank car, BLT CER 1903 as #13, REN 090, Dupont TRK, 2WH #49 and 1WH #60L type S.K. 15 hp MOT, K-10 CTL, 8-9-29 Highway Oil Co removed tank at E. 55th Station, SCR at E. 34th 1-20-36

0221  flusher, ST, FN 091, BLT by CER in 1903 and SCR in 1929

0222  flusher, ST, LEN 21', iron tank car, BLT 1907 Miller & Noblock as #4, REN 076 and later REN 092, Dupont TRK, 2WH #49 and 1WH #60L type S.K. 15 hp MOT, K-10 CTL, car had tank capacity of 3,000 gallons of water and W 26,000 lbs, SCR 3-8-41 (also listed as SCR 3-26-41)

0223  flusher, LEN 29' 6", iron tank car, BLT CR 1915, two Brill 27F-1 TRKS, 4WH #101 and 1WH #60L pump MOT, K-12 CTL, 5,000 gallon water tank and W 36,000 lbs, SCR 9-19-51

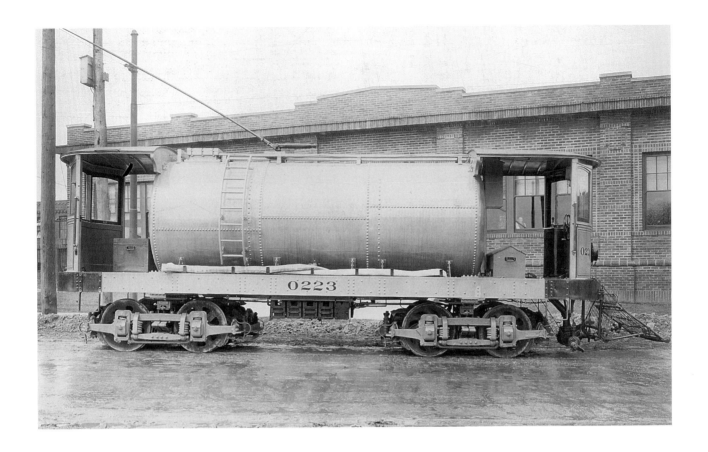

0223, flusher (iron tank car), Cleveland Railway, 1915. *(Cleveland Railway photo, Blaine Hays Collection)*

0224    flusher, LEN 29' 6", iron tank car, BLT CR 1915, two Brill 27F-1 TRKS, 4WH #101 and 1WH #60L pump MOT, K-12 CTL, 5,000-gallon water tank and W 36,000 lbs and was used in Lakewood to fulfill street flushing franchise obligations until well after World War II, operated in "Parade of Progress" 4-27-52, SCR 11-9-53

0231    yard locomotive, ST, LEN 19' 8", BLT CR, D 4-20-14, Dorner & Dutton TRK, 2WH #101B-6 MOT, K-10 CTL, W 26,500 lbs, assigned to Madison then E. 55th stations, SCR 8-18-53

0232    yard locomotive, ST, LEN 19' 8", BLT CR, D 4-20-14, Dorner & Dutton TRK, 2WH #101 MOT, K-10 CTL, W 26,500 lbs, assigned to St. Clair Station, SCR either 3-21-41 or 3-27-43 (also listed as SCR 3-28-41!)

0233    yard locomotive, ST, LEN 19' 8", BLT CR, D 5-27-14, Dorner & Dutton TRK, 2WH #49 MOT, two K-10 CTL, W 20,900 lbs, assigned to Windermere Car House, SCR 11-2-49

0234    yard locomotive, ST, LEN 19' 8", BLT CR, D 7-23-14, Dorner & Dutton TRK, 2WH #101B-6 MOT, two K-10 CTL, W 26,500 lbs, assigned to Superior Station, SCR 4-28-52

0235    yard locomotive, ST, LEN 19' 8", BLT CR, D 7-23-14, Dorner & Dutton TRK, 2WH #49 MOT, two K-10 CTL, W 20,900 lbs, assigned to Denison Station, SCR 7-10-47

0236 yard locomotive, ST, LEN 19' 8", BLT CR, D 7-23-14, Dorner & Dutton TRK, 2WH #49 MOT, two K-10 CTL, W 20,900 lbs, assigned to Brooklyn Station, SCR 3-15-49

0237 yard locomotive, ST, len 19' 8", BLT CR, D 11-28-14, Dorner & Dutton TRK, 2WH #49 MOT, two K-10 CTL, W 20,900 lbs, assigned first to Miles and then to E. 55th stations, SCR 10-6-47

0238 yard locomotive, ST, LEN 19' 8", BLT CR, D 11-28-14, Dorner & Dutton TRK, 2WH #101 MOT, two K-10 CTL, W 26,500 lbs, assigned to St. Clair Station and in later years used in salt storage service, SCR 4-7-48

0239 yard locomotive, ST, BLT by CR, D 11-28-14, Dorner & Dutton TRK, 2 WH #49 MOT, two K-10 CTL, W 20,900 lbs, assigned to Harvard Shops, SCR 3-19-41 (also listed as SCR 3-20-41)

0240 yard locomotive, ST, LEN 19' 8", BLT CR, D 11-28-14, Dorner & Dutton TRK, 2WH #49 MOT, two K-10 CTL, W 20,900 lbs, assigned to Windermere Station, SCR 4-28-52

0241 yard locomotive, ST, LEN 19' 9", BLT CR, D 11-16-15, Dorner & Dutton TRK, 2WH #49 MOT, two K-10 CTL, W 20,900 lbs, assigned to St. Clair Station, SCR 11-2-49

0242 yard locomotive, ST, LEN 19' 9", BLT CR, D 11-27-15, Dorner & Dutton TRK, 2WH #49 MOT, two K-10 CTL, W 20,900 lbs, assigned to Brooklyn Station, SCR 2-9-50

0243 yard locomotive, ST, LEN 19' 9", BLT CR, D 11-27-15, Dorner & Dutton TRK, 2WH #101 MOT, two K-10 CTL, assigned to Harvard Shops, SCR 3-28-41 (also listed as SCR 3-29-41)

0244 yard locomotive, ST, LEN 19', BLT CR, D 2-16-23, Dorner & Dutton TRK, 2 WH #49 MOT, two K-10 CTL, first assigned to Superior Station and then to St. Clair Station, used in salt storage service in later years, SCR 4-7-48

0245 yard locomotive, ST, LEN 19', BLT CR, D 2-16-23, Dorner & Dutton TRK, 2WH #101 MOT, two K-10 CTL, W 26,500 lbs, assigned to Windermere Station, SCR 11-2-49

0248 yard locomotive, ST, LEN 12' 4", BLT CER 1906 as #0157, Brill 21 TRK, 2WH #49 MOT, K-10 CTL, assigned to the Viaduct Powerhouse until its closure 11-6-25, transferred to Harvard Shops, SCR probably 1951

0249 yard locomotive, ST, LEN 14' 4", BLT Van Dorn Iron Works Co. 1902 as #2, later REN 0158, Dupont TRK, 2WH #49 MOT, K-10 CTL, assigned to the Viaduct Powerhouse until closure on 11-6-25, transferred to Harvard Shops, SCR 11-26-52

0500 dump freight, LEN 40' 6", BLT Differential Steel Car, D 12-19-20, Standard TRKS, 4WH340 MOT, group HL CTL 264T2 15B master, W 47,800 lbs, SCR 6-4-48

0501     dump freight, LEN 40' 6", BLT Differential Steel Car, D 12-20-20, Standard TRKS, 4WH340 MOT, HL CTL, W 47,900 lbs, SCR 10-4-51

0502     dump freight, LEN 40' 6", BLT Differential Steel Car, D 10-20-20, Standard TRKS, 4WH340 MOT, HL CTL, W 48,400 lbs, SCR 11-2-48

0503     dump freight, LEN 40' 6", BLT Differential Steel Car, D 10-24-20, Standard TRKS, 4WH340 MOT, HL CTL, W 48,100 lbs, SCR 5-16-51

0504     dump freight, LEN 40' 6", BLT Differential Steel Car, D 7-22-22, Standard TRKS, 4WH340 MOT, HL CTL, W 47,400 lbs, SCR in 1948

0505     dump freight, LEN 40' 6", BLT Differential Steel Car, D 7-22-22, Standard TRKS, 4WH340 MOT, HL CTL, assigned first to St. Clair Station then to Windermere Station, W 47,900 lbs, SCR 3-9-51

0506     dump freight, LEN 40' 6", BLT Differential Steel Car, D 7-28-22, Standard TRKS, 4WH340 MOT, HL CTL, W 47,400 lbs, SCR probably in 1951

0507     dump freight, LEN 40' 6", BLT Differential Steel Car, D 7-28-22, Standard TRKS, 4WH340 MOT, HL CTL, W 47,100 LBS, SCR 9-12-51

0508     dump freight, LEN 40' 8", BLT Differential Steel Car, D 5-25-23, Standard TRKS, 4WH340 MOT, HL CTL, W 46,900 lbs, SCR 9-17-51

0509     dump freight, LEN 40' 6", BLT Differential Steel Car, D 5-25-23, Standard TRKS, 4WH340 MOT, HL CTL, W 48,000 lbs, SCR 9-16-48

0510     dump freight, LEN 40' 6", BLT Differential Steel Car, D 5-23-23, Standard TRKS, 4WH340 MOT, HL CTL, W 46,900 lbs, SCR probably in 1946

0511     dump freight, LEN 40' 6", BLT Differential Steel Car, D 5-23-23, Standard TRKS, 4WH340 MOT, HL CTL, W 46,900 lbs, SCR 12-26-52

0512     dump freight, LEN 39' 10", BLT Differential Steel Car, D 5-25-25, Differential TRKS, 4WH340 MOT, HL CTL, W 48,800 lbs, SCR 9-18-47

0513     dump freight, LEN 39' 10", BLT Differential Steel Car, D 5-25-25, Differential TRKS, 4WH340 MOT, 1WH dump motor with 7 hp, HL CTL, W 48,800 lbs, assigned to St. Clair Station, SCR 9-26-51

0514(1) dump freight, LEN 39' 10", BLT Differential Steel Car, D 6-12-25, Differential TRKS, 4WH340 MOT, HL CTL, W 48,800 lbs, SCR in 1951

0515     dump freight, LEN 39' 10", BLT Differential Steel Car, Differential TRKS, 4WH340 MOT, HL CTL, W 48,800 lbs, SCR 9-21-51

0516    dump freight, LEN 40' 5", BLT Differential Steel Car, D 9-1927, Differential TRKS, 4WH340P MOT, HL CTL, W 50,200 lbs, assigned to Superior Station, REB 1953 at Harvard for use on CTS Rapid Transit, car operated in "Parade of Progress" pulling flat car 0611 on 4-27-52, sold by RTA to NORM 1979, car SCR by museum at Kingsbury Shops 2-1979, parts salvaged

0517    dump freight, LEN 40' 5", BLT Differential Steel Car, D 9-1927, Differential TRKS, 4WH340P MOT, HL CTL, W 50,500 lbs, assigned to Brooklyn Station, dismantled at Harvard Yards and major components sold 12-10-53

0518    dump freight, LEN 40' 5", BLT Differential Steel Car, D 9-1927, Differential TRKS, 4WH340 MOT, HL CTL, W 50,300 lbs, assigned to St. Clair Station, REB at Harvard Shops 1953 for use on CTS Rapid Transit, operated in "Parade of Progress" 4-27-52, sold to NORM 1979, moved to museum from Kingsbury Shop 12-1979, now on display as artifact at museum

0519    dump freight, LEN 40' 5", BLT Differential Steel Car, D 9-1927, Differential TRKS, 4WH340 MOT, HL CTL, W 50,200 lbs, assigned to Rocky River Station, dismantled at Harvard Yards, major components sold 12-10-53

0520    dump freight, LEN 40' 5", BLT Differential Steel Car, D 9-1927, Differential TRKS, 4WH340 MOT, HL CTL, W 50,200 lbs, SCR summer 1953

0521    dump freight, LEN 40' 5", BLT Differential Steel Car, D 9-1927, Differential TRKS, 4WH340 MOT, HL CTL, W 50,200 lbs, SCR 8-18-53

0522    dump freight, LEN 40' 5", BLT Differential Steel Car, D 1927, Differential TRKS, 4WH340 MOT, HL CTL, W 50,200 lbs, SCR summer 1953

0523    dump freight, LEN 40' 5", BLT Differential Steel Car, D 9-1927, Differential TRKS, 4WH340 MOT, HL CTL, W 50,200 lbs, SCR 1-21-54

0524    dump freight, len 35' 6", BLT Differential Steel Car, D 7-7-15, Standard TRKS, 4WH340 MOT, Master B15 HL CTL 264T2, W 46,600 lbs, SCR in 1946?

0570    dump freight trailer, LEN 34' 4", BLT Differential Steel Car, D 7-7-15, FN 027, Standard TRKS, 2WH340 MOT, HL CTL, W 39,700 lbs, SCR 3-27-41 (also listed as SCR 3-28-41)

0571    dump freight trailer, LEN 34' 4", BLT Differential Steel Car, D 7-7-15, FN 028, Standard TRKS, 2WH340 MOT, HL CTL, W 39,700 lbs, SCR 3-13-41 (also listed as SCR 3-24-41)

0572    dump freight trailer, LEN 40' 6", BLT Differential Steel Car, D 5-25-23, FN 090, Standard TRKS, W 32,400 lbs, SCR 3-16-50

0573    dump freight trailer, LEN 40' 6", BLT Differential Steel Car,

0523, dump freight, Differential Steel Car, 1927. *(Cleveland Railway photo, Blaine Hays Collection)*

        D 5-26-23, FN 091, Standard TRKS, W 32,400 lbs, SCR 7-10-51

0574    dump freight trailer, LEN 35' 6", BLT Differential Steel Car, D 8-20-22, FN 092, Standard TRKS, W 32,500 lbs, SCR 1948

0575    dump freight trailer, LEN 35' 6", BLT Differential Steel Car, D 8-20-22, FN 093, Standard TRKS, W 32,400 lbs, SCR 1948

0576    dump freight trailer, LEN 40' 8", BLT Differential Steel Car, D 1920, FN 094, Standard TRKS, W 33,300 lbs, SCR 11-15-48

0577    dump freight trailer, LEN 40' 6", BLT Differential Steel Car, D 1920, FN 095, Standard TRKS, W 33,400 lbs, SCR 3-16-50

0578    dump freight trailer, LEN 39' 10", BLT Differential Steel Car, D 5-25-25, FN 043, Differential TRKS, 1WH dump 7 hp MOT, W 35,400 lbs, SCR 9-12-51

0579    dump freight trailer, LEN 39' 10", BLT Differential Steel Car, D 6-12-25, FN 045, Differential TRKS, 1WH dump 7 hp MOT, W 35,400 lbs, SCR 1951

0580    dump freight trailer, LEN 39' 9", BLT Differential Steel Car, D 10-24-27, FN 033, Differential TRKS, W 36,600 lbs, SCR 8-27-53

0581    dump freight trailer, LEN 39' 9", BLT Differential Steel Car, D 10-24-27, FN 034, Differential TRKS, W 33,500 lbs, SCR in 1951

043, Thew electric shovel and dump freight trailer, Differential Steel Car, 1925. (*Cleveland Railway photo, Blaine Hays Collection*)

0582    dump freight trailer, LEN 39' 9", BLT Differential Steel Car, D 11-11-27, FN 035, Differential TRKS, W 33,500 lbs, SCR in early 1950s

0583    dump freight trailer, LEN 39' 9", BLT Differential Steel Car, D 12-1-27, FN 036, Differential TRKS, W 33,500 lbs, SCR 8-27-53

0610    flat car, LEN 38' 1", BLT Differential Steel Car, D 11-3-28, Differential TRKS, W 25,500 lbs, REB for use on SHRT and was shipped on 11-4-53, car sold to NORM 1981, shipped to museum with Crane Car 0710 and displayed

0611    flat car, LEN 38' 1", BLT Differential Steel Car, D 11-3-28, Differential TRKS, W 25,500 lbs, pulled by 0516 in "Parade of Progress" 4-27-52, REB at Harvard Shops 1953 for use on CTS Rapid Transit, sold to NORM 1981, shipped to museum

0612    flat car, LEN 38' 1", BLT Differential Steel Car, D 11-3-28, Differential TRKS, W 25,500 lbs, SCR 2-10-54

0613    flat car, LEN 38' 1", BLT Differential Steel Car, D 11-3-28, Differential TRKS, W 25,500 lbs, SCR 2-10-54

0614  flat car, LEN 38' 1", BLT Differential Steel Car, D 11-3-28, Differential TRKS, W 25,500 lbs, SCR 2-10-54

0615  flat car, LEN 38' 1", BLT Differential Steel Car Co, D 11-3-28, Differential TRKS, W 25,500 lbs, transferred to CTS Rapid Transit 1953, sold to the NORM 1981, transferred to museum where it is currently on display

0616  flat car, see "rapid transit" section

0617  flat car, see "rapid transit" section

0618  dump freight, see "rapid transit" section

0619  hopper car, see "rapid transit" section

619  fare box storage, former passenger car became fare box container at St. Clair Station in 1924, SCR 1935

0631  dump freight, LEN 36' 4", BLT Orenstein Arthur Kopple 1913 as #017, triple bed dump car, Standard TRKS, 4WH #305 MOT, 15B Master HL CTL, W 54,000 lbs, probably SCR 7-3-51

0633  dump freight, LEN 36' 4", BLT by Orenstein Arthur Kopple 1913 as #019, triple bed dump car, Baldwin 4923 TRKS, 4WH #305 MOT, 15B Master HL CTL, W 61,200 lbs, probably SCR 7-3-51

0640  trailer, see FN 2203 for details

0641  trailer, see FN 2280 for details

0642  trailer, see FN 2214 for details

0643  trailer, see FN 2282 for details

0644  trailer, see FN 2285 for details

0647  CR T 16 "Ten-Hundred," see FN 1010 for details

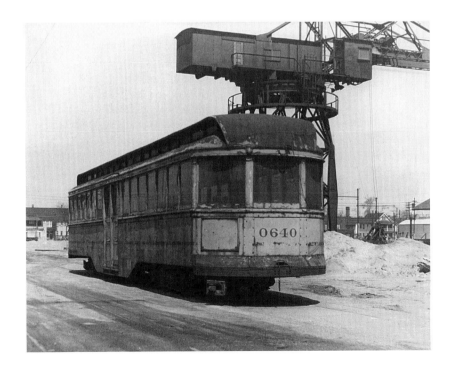

0640, section headquarters trailer shown with Harvard Shop gantry crane, G. C. Kuhlman, 1913, CR type T 4. *(Jim Spangler photo)*

0648    CRT 16 "Ten-Hundred," see FN 1005 for details

0649    CRT 16 "Ten-Hundred," see FN 1000 for details

0650    CRT 16 "Ten-Hundred," see FN 1045 for details

0651    flat car, ST, LEN 19', BLT CER 1901 as #14, REN 039, Dorner & Dutton TRK, W 11,640 lbs, SCR at E. 34th 1-18-36

0653    flat car, ST, LEN 20', BLT CER 1896 as #5, REN 041, Dupont TRK, W 12,000 lbs, SCR 12-6-45

0656    flat car, ST, LEN 20', BLT CER 1896 as #19, REN 048, Dorner & Dutton TRK, W 11,000 lbs, SCR 3-26-41

0665    flat car, ST, LEN 19', BLT W.B. 1909 as #062, Dorner & Dutton TRK, W 10,900 lbs, SCR at E. 34th 1-18-36

0672    flat car, ST, len 19', BLT CR 2-1913 as #071, Dupont TRK, W 13,000 lbs, SCR 5-2-47

0676    flat car, ST, LEN 19', BLT CR 1913 as #075, Dupont TRK, W 12,500 lbs, stored at Harvard Shops in later years pending repairs, repair cost of $500.00 was not warranted 8-4-38, SCR 3-21-41 (also listed as SCR 3-25-41)

0680    flat car, LEN 30', BLT CR, D 1-2-24 as #046, Brill 51E-1 TRKS, W 22,100 lbs probably SCR on 7-3-51

### Flat Car Renumberings, Recapitulation

*Asterisk indicates units listed in roster under a four-digit number.*

| | |
|---|---|
| 0650 | FN 037 |
| 0658 | FN 050 |
| 0666 | FN 063 |
| 0674 | FN 073 |
| ★0651 | FN 039 |
| 0659 | FN 051 |
| 0667 | FN 065 |
| 0675 | FN 074 |
| 0652 | FN 040 |
| 0660 | FN 052 |
| 0668 | FN 066 |
| ★0676 | FN 075 |
| ★0653 | FN 041 |
| 0661 | FN 056 |
| 0669 | FN 068 |
| 0677 | FN 096 |
| 0654 | FN 044 |
| 0662 | FN 057 |
| 0670 | FN 069 |
| 0678 | FN 099 |
| 0655 | FN 047 |
| 0663 | FN 058 |

| | |
|---|---|
| 0671 | FN 070 |
| 0679 | FN 0100 |
| ★0656 | FN 048 |
| 0664 | FN 059 |
| ★0672 | FN 071 |
| ★0680 | FN 046 |
| 0657 | FN 049 |
| ★0665 | FN 062 |
| 0673 | FN 072 |
| 0690 | welder, LEN 41' 8", (28' box), BLT Niles 1903 as passenger car 109, REN 01 in 1911 as welder and 0690 in 1922, Brill 27F TRKS, 4WH #101 MOT, K-12 CTL, burned beyond repair while on job at St. Clair & Ontario 2-1-37, SCR 7-21-37 |
| 0691 | welder, LEN 42', (28' box), BLT Niles (card file says Kuhlman?) 1902 as #110, REN 05 1921 when CON to welder, REN 0691 1930, Brill 27F TRKS, 4WH #101 MOT, K-12 CTL, SCR 4-24-41 (also listed as SCR 4-28-41) |
| 0692 | welder, FN passenger 358, CON 1922, SCR 1936 |
| 0693 | welder, LEN 41' 3" (28' box), BLT Brill 1903 as passenger 368, REN 10, REN 030 when CON to welder 1913, 3-20-30 REN to 0693, Brill 27F TRKS, 4WH #101 MOT, K-12 CTL, SCR 1947 |

0691, welder, Niles, 1902, as passenger 110. *(Karel Libenauer photo, Northern Ohio Railway Museum)*

0694 welder, LEN 41' 3", (28' box), BLT Brill 1899 as passenger 370, REN 11, REN 024 when CON to welder, REN 0694 1915, Brill 27F TRKS, 4WH #101B-6 MOT, K-12 CTL and SCR 11-21-47

0695 welder, LEN 43' 6", BLT Brill 1903 as passenger 198, REN 0695 1910 when CON to welder, Brill 27F TRKS, 4WH #101 MOT, K-6 CTL, SCR 6-7-50

0696 welder, LEN 43' 6", BLT Kuhlman 1902 as passenger 648, REN 0696 1929 when CON to welder, Brill 27F TRKS, 4WH #101 MOT, K-6 CTL, SCR 9-11-47

0697 welder, LEN 43' 6", BLT Brill 1903 as passenger 116, REN 0697 1931 when CON to welder, Brill 27F TRKS, 4WH #101 MOT, K-6 CTL, SCR 4-15-41 (also listed as SCR 4-17-41)

0698 welder, LEN 43' 6", BLT Brill 1903 as passenger 117, REN 0698 1931 when CON to welder, Brill 27F TRKS, 4WH #101 MOT, K-6 CTL, SCR 4-28-41 (also listed as SCR 4-30-41)

0699 welder, LEN 41' 3", (28' box), BLT Brill 1900 as passenger 741, REN 0699 1928 when CON to welder, Brill 27F TRKS, 4WH #101 MOT, K35G CTL, SCR 4-28-41 (also listed as SCR 4-29-41)

0710 crane car, LEN 43', BLT Differential Steel Car, D 9-1925 as #0156, Standard TRKS, 4WH variable speed 7 hp and 4WH340 MOT, HL CTL, REN 11-1953 for SHRT at Harvard and shipped to Shaker 2-1-54, crane removed after main gear sheared 1978, became motorized flat, sold to NORM 1981, moved to museum, now exhibit at Westfield Township museum in Medina County

0711 crane car, LEN 43', BLT Differential Steel Car, D 9-1925 as #0157, Standard TRKS, 4WH variable speed 7 hp and 4WH340 MOT, HL CTL, REN for CTS Rapid Transit, transferred from Harvard 3-10-53, sold NORM 1980, transferred to museum where it is currently an exhibit

0712 crane car, LEN 43', BLT Differential Steel Car, D 10-17-25 as # 0158, Standard TRKS, 4 variable speed 7 hp and 4WH340 MOT, HL CTL, saved for use on CTS Rapid Transit but found in worse condition than 0711 so was replaced by it, SCR 6-2-54

0713 crane car, LEN 43', BLT Differential Steel Car, D 10-17-25 as #0159, Standard TRKS, 4 variable speed 7 hp and 4WH340 MOT, HL CTL, SCR 11-20-53

0730 flat car, LEN 64' 1¾", BLT Differential Steel Car, D 5-1925 first as number 036, REN 088, then REN 089, Differential TRKS, W 30,500 lbs, SCR 4-28-54

0731 flat car, LEN 64' 1¾", BLT Differential Steel Car, D 6-12-25, either Standard or Differential TRKS?, W 30,500 lbs, SCR 5-7-54

0740 derrick car, LEN 48', BLT Van Dorn Iron Works, D 3-14-13

as #0153, Baldwin 427 TRKS, 4WH #93 and 1 GE 40 hp MOT, two K-6 CTL, W 84,400 lbs, among the heaviest cars on CR, used during track reconstruction in the street to pull pavement plows, SCR 9-5-51

0741 derrick car, LEN 48' 10", BLT Brown Hoist Crane, D 4-16-14 as #0154, Baldwin 427 TRKS, 1 GE type C.O. 40 hp and 4WH #305 MOT, two K-6 CTL, W 84,400 lbs, leased to receivers of Lake Shore Electric Railway in 1938 to lift bridge girders from Berlin Heights, Bay Village (Huntington), and Cahoon Park trestles SCR 1-16-53

0760 flat motor car, LEN 35', BLT Miller & Noblock as #9 1900, REB by CR 1910 as #34, REB again by V D & D 1914 as Sprinkler Car 013, known to CR employees as "Jumbo," was heaviest car on CR, 85,100 lbs, Baldwin 427 TRKS, 4WH #101 MOT, 2K-6-1 Speed Regulator CTL, SCR 6-10-54

0730, flat car, Differential Steel Car, 1925. *(Cleveland Railway photo, Blaine Hays Collection)*

PART A: STREETCAR ROSTER · 151

0741, derrick, Brown Hoist, 1914.
(*Jim Spangler photo*)

0761    flat motor car, LEN 35', BLT CR, D 1-20-13 as #3, REN 015, used to switch 1927 Cleveland Industrial Exhibition displays into Public Hall, one of car's greatest feats was pulling a 318,400-lb Nickel Plate Hudson steam locomotive, number 173, into exhibit grounds! Standard TRKS, 4WH #307F MOT, K35G Speed Regulator CTL, performed interchange with the Newburgh & South Shore Railroad at Harvard Yards, SCR 2-10-54

0765    sand car, (tank) LEN 30', BLT CR, D 5-10-19 as #0176, traveled over entire system, to all stations and loops, to replenish sand in all containers, car Brill 27F TRKS, 4WH #101 MOT, K35G CTL, SCR 5-4-53

0775    Crane car, LEN 22' (boom 50'), BLT Ohio Crane, D 11-27-17 as #0155, aside from two funeral cars 0775 was only other CR car painted black, incorrectly listed by Morse in his roster as "0725" and incorrectly listed by Christiansen in his roster as "formerly 0165," causing considerable confusion, car came with factory #3269 crane, W 64,000 lbs, standard Master Car Builders (railroad) trucks, 1WH #70 type M.C. MOT, W.H.R. 164A CTL, car sold to Summer & Co. scrap iron dealer on Whiskey Island 7-1-54 and is believed to be extant

0800    rail scrubber, LEN 28', BLT American Car 1919, D 7-24-34, was a Birney Car numbered 371 on Penn-Ohio interurban system (Youngstown), W 29,300 lbs empty and 33,300 lbs with water in two 250-gallon tanks, Brill 79E TRK, 2WH #101 MOT, two K-10 CTL, two 10 hp grinding motors that ran at

152 · CLEVELAND'S TRANSIT VEHICLES

0761, flat motor car, Cleveland Railway, 1913. *(Cleveland Railway photo, Blaine Hays Collection)*

0765, sand car, Cleveland Railway, 1919. *(Jim Spangler photo)*

0800, rail scrubber, American Car, 1919, as Youngstown city car 371. *(Jim Spangler photo)*

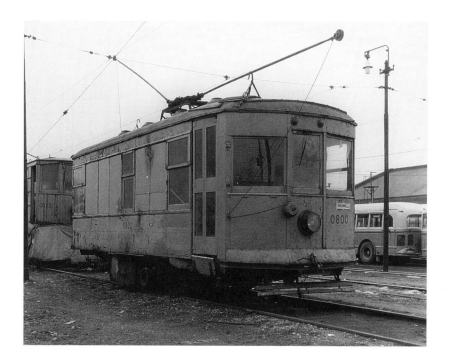

600 rpm transferred to CTS Rapid Transit construction in 1953, moved to Shaker's Kingsbury Shops. Major portions of car, including trucks, controls, were sold to Illinois Railway Museum 8-2-57 to provide power for Illinois Terminal Railway Birney number 170

815     former passenger car became South Brooklyn (Brooklyn Station) waiting room in 1924

### Miscellaneous Way Department Equipment—Street Railway

0905, 0906, 0907, 0908, 0909 • Thew Electric Shovel: 0905, BLT 1913 (FN0252), SCR 10-4-45; 0906, BLT 1913, FN 0253, SCR 10-4-45; 0907 BLT 1914, SCR 10-4-45; 0908 and 0909 sold 6-7-53

0920, 0921, 0922, 0923 • Clark Concrete Breakers (pavement pounders): 0920 BLT 1923, FN 3; 0921 BLT 1926, FN 01885W; 0922 BLT 1929; 0923 (concrete cutter), D 10-14-48 along with carrier 03

0930–0931 • Clark Pavement Plows: 0930 BLT Union Engineering Co. (Beverage Engineering), D 12-9-19; 0942–0943 Springfield Gasoline Rollers, 0943 sold 6-19-56

Fordson Schramm Air Compressors with Snow Brooms

| | | | |
|---|---|---|---|
| tractor | 0950 | compressor | 01350 |
| | 0951 | | 01353, 0951, 01353 SCR 1949 |
| tractor | 0952 | compressor | 01352, 0952, 01352 SCR 1949 |

Fordson-Schramm Air Compressors
- tractor 0953 compressor 01351
- 0954 01356
- 0955 01354
- 0956 01357

Fordson-Metalweld Air Compressors, all SCR 1949
- 0957 01355

Warco Graders (without compressors)
- tractor 0958 compressor 0962
- 0959 01358
- 0960 01361, sold 5-7-53
- 0961 01360
- 0963 01359

(McCormick-Metalweld Air Compressors)

0979 • Lull Tractor Mounted Loader, D 1947
- tractor 0964 compressor 01362, sold 5-7-53
- 0965 01363
- 0966 01364, sold 5-7-53
- 0967 01365

(McCormick-Davey Air Compressors)     0967, sold 5-7-53
- motor 0968 compressor 1368
- 0969 1369
- 0970 1370
- 0971 1371
- 0972 1372

(Davey Portable Air Compressors)   0976 D 1947

0973, 0974, 0975 • Le Roi Air Compressors, D 1947

0977, 0978, 0979 • Case Tractor Mounted Snow Sweepers, D 1947, 0979 D 1949

0981, 0982, 0983, 0984 • Davey Air Compressors, D 1947, 0982 sold 2-3-58

0985 Hough Loader, D 1947

0986, 0987, 0988 • Lull Loaders, 0986, 0987 D on 10-1-48, 0988 D 1949

0990, 0991, 0993, 0994, 0998 (sold 3-12-53), 0999 (SCR 10-16-53), 01000, 01001 • Koehring & Rex Concrete Mixers: 0990 FN 0260; 0991 FN 0262 (1913); 0992 FN 0263; 0993 FN 0264; 0994 FN 0265; 0995 FN 0267; 0996 FN E-10169; 0997 FN E-10170

0992–0995 • Screed Koehring Mixers, modified

0996, 0997 • Rex Concrete Mixers, 0997 SCR 1941

01010, 01011, 01012, 01013 • Union Engineering Co. Conveyors, SCR 1941; 01010 FN 0270 SCR 3-10-41; 01011 FN 0268 SCR 3-10-41; 01012 FN 0266 SCR 3-10-41; 01013 FN 01857W SCR 3-10-41

01020 (SCR 7-28-53), 01021 (SCR 10-7-52), 01022 (SCR 10-7-52), 01023 • Salt Loaders: 01020 BLT 1928; 01021 BLT 1930; 01022 BLT 1942; 01023 D 1949

01030, 01031, 01032, 01033 • Concrete Mixers, FN respectively 1, 2, 3, 0261

01034, 01035, 01036, 01037 • Smith Mixers, BLT 1930, SCR 1951 (01035 was grouter)

01050, 01051 • Kerwin Track Grinders, had K-10 WH CTL, 01050 BLT 1910; 01051 BLT 1911, both SCR 11-20-53

01053   Vulcan Track Grinder, BLT 1927

01054, 01055, 01056, 01057 • Hercules Track Grinders, BLT 1927

01058   Universal Track Grinder, BLT 1927

01059   Imperial Automatic Track Grinder, BLT 1927, SCR 4-6-54

01060, 01061 (SCR 8-18-53), 01062, 01063 • Atlas Track Grinders, BLT 1927

01064, 01065, 01066, 01067 • Vulcan Track Scrubbers, BLT 1928 to 1932, last three SCR 1951

01069, 01070 • Biax Track Grinders (flexible)

01071, 01072 • Hercules Track Grinders

01073, 01074, 01075, 01076, 01077, 01078 • Mall Flexible Shaft Grinders, last four D 1942

01079, 01080, 01081, 01082 • Mall Rail Grinders, D in 1947

01150, 01151 • Tool Carrying Cars, hand cars, both SCR 3-26-54

01366, 01367 • Davey Air Compressors, 01366 on truck T-68 (sold 5-17-56) and 01367 on truck T-69 (sold 5-17-56)

01400, 01401, 01402, 01403, 01404, 01405 (SCR 8-18-53), 01406, 01408, 01409, 01410 (SCR 5-17-56), 01411, 01412 (SCR 5-17-56), 01413 (SCR 5-17-56) • Una Welding Co. Dynamotors (electric welders), 01411 SCR 1954

01407, 01408 • Lincoln Dynamotor (elect welder) 01408 SCR 1-29-58

01414, 01415 • Lincoln Dynamotors; truck mounted, 01414 on truck T-68 and 01415 on truck T-69

01416, 01417, 01418 • General Electric Welders, D 1947

01419, 01420 (sold 8-8-58), 01421, 01422 (sold 8-8-58), 01423 (sold 8-58), 01424 (sold 25 SCR 2-11-54), 01427, 01428, 01429, 01432 (SCR 5-17-56), 01433 (SCR 5-17-56), 01434 (SCR 5-17-56) • Lincoln Electric Welders, D 1947

Birch Excel Sand Spreaders
    serial 4201 • model MS
    serial 4202 • model MS

01425, 01426 • Equipment Trailers R-05; Bonding Trailer

01430, 01431 • Ingersoll-Rand Tie Tamper Compressors

01471, 01472, 01473, 01474, 01475, 01476, 01477, 01478, 01479, 01480 • Chicago Pneumatic Co. track drills

01490   W. H. Keller track drill

01492   Ingersoll-Rand track drill

01493, 01494, 01495, 01496, 01500, 01501 • Duntley track drills

01500(2)   Austin Western Grader, D in 1949

01571   Acetylene Generator, D in 1949
01650, 01651, 01652, 01653, 01654, 01655 • Pre-heaters
01707, 01708, 01709, 01710, 01712, 01713, 01715 (sold 2-3-58) •
    Littleford Tar Kettles, 01715, D 1-29-48
01714   Blankner Joint Filling Machine, sold 2-3-58
T-88, T-89 • Tool Trailers
01-02   Pole Trailers, D 1-29-48
T-90, T-91, T-92, T-93, T-94, T-95, T-96, T-97, T-98, T-99 • REN S-99;
    Bus Trailers (original numbers unknown)
S-99   Used by Transportation Department

### Postwar Service Cars: Street Railway

As World War II ended, many service cars were old and unreliable, hundreds of passenger cars and trailers were being discontinued, and CRT 16 Ten-Hundred cars were being replaced by PCC cars. With a huge street railway system still in place, it was decided to convert some units for work service. Specifically, the line cars were especially old and needed replacing. Since the ancient single-truck snow sweepers constantly derailed and caused much trouble, they too needed to be replaced. A car to tow the Irwin Grinder was also needed, and so a new, small work fleet evolved. The first two cars in the following roster were inspired by CTS general manager Donald C. Hyde, as he came to CTS from the Milwaukee system.

0136(2)   snow sweeper, DT, former Milwaukee, WI #B-33, purchased 1951, REN in Cleveland to 01036, indicating that it was intended to be included in miscellaneous way equipment, later REN into service car fleet as 0136, served St. Clair Station 1952, moved to Denison Station, SCR there 9-14-56

0137(2)   snow sweeper, DT, former Milwaukee, WI #B-35, purchased 1951, served Denison Station 1952, moved to Harvard Shops after streetcars ended on 1-24-54, REB for use on CTS Rapid Transit, transferred to Windermere 1953. After years of idleness and a fire in 1959, car was retired 9-22-60, parts sold to Connecticut Electric Railway 10-4-60.

0173   derrick, origin unknown, began appearing on CTS listings 1947, whether car was a REN crane is not known, SCR 1-13-54

0208(2)   tow car, FN 208, CON 1950 with seats removed and door cut in right rear, car towed the Irwin Grinder, operated in "Parade of Progress" 4-27-52, SCR 10-7-53

0136 (2), double-truck Milwaukee snow sweeper, still numbered 01036 for miscellaneous Way Department equipment. *(Cleveland Transit photo, Blaine Hays Collection)*

| | |
|---|---|
| 0363 | line car, FN 363, CON 1950 by removing seats and adding a flat roof platform, had hand-lettered inside front door "Price's Power Car—pride of line dept," in reference to Clarence "Tiny" Price, a popular line department employee and motorman, operated in "Parade of Progress" 4-27-52, SCR 3-30-54 |
| 0514(2) | flat car, ST, REB from horsecar frame, used to the end (1954), however does not show up in inventory |
| 0640 | section headquarters trailer, FN 2203, CON 1947, SCR 3-29-54 |
| 0641 | section headquarters trailer, FN 2280, CON 1947, SCR 1-14-54 |

0363, FN 363, converted to line car, 1950. *(Jim Spangler photo)*

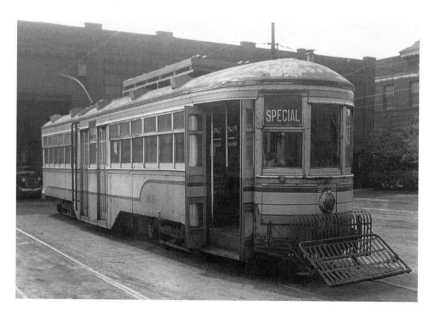

158 · CLEVELAND'S TRANSIT VEHICLES

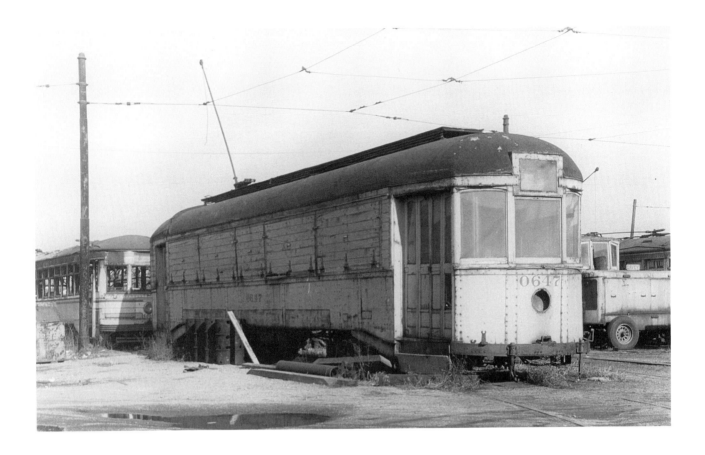

0647, section headquarters car, FN 1010, G. C. Kuhlman, 1913. (*Jim Spangler photo*)

| | |
|---|---|
| 0642 | section headquarters trailer, FN 2214, CON 1947, SCR 1-13-54 |
| 0643 | section headquarters trailer, FN 2282, CON 1947, SCR 12-2-52 |
| 0644 | section headquarters trailer, FN 2285, CON 1947, SCR 5-5-54 |
| 0647 | section headquarters car, FN 1010, CON 1947, SCR 3-29-54 |
| 0648 | section headquarters car, FN 1005, CON 1947, SCR 3-22-54 |
| 0649 | section headquarters car, FN 1000, CON 1947, SCR 2-26-53 |
| 0650 | section headquarters car, FN 1045, CON 1947, SCR 5-17-54 |

ROSTER SHORTS

1. *The Peter Smith Coal Stove:* An ancillary feature of Cleveland's streetcars was the use of anthracite-coal-fired Peter Smith car stoves rather than electric heat. The stove had only a small 600-volt DC blower motor. With a two-man crew, the conductor tended the fire. Electric heat would use a large amount of energy and would add to demand. The stove was a low-cost solution. Each fall, a dump freight work car would deliver a load of "nut"-size anthracite coal to each carbarn. Coal stoves lasted until August 1951.

The Peter Smith coal stove, used on many streetcars until 1951, kept passengers tolerably warm during winter months. *(Blaine Hays Collection)*

2. *The "Cowcatcher" or Eclipse Fender:* The elaborate iron grate that adorned a large number of streetcars was the invention of Clevelander Benjamin Lev and was patented on 10-20-04. Mr. Lev was able to sell his invention to streetcar officials by successively allowing a car equipped with the device to hit him at increasing speeds. Each time he arose unhurt. The flat front grate would swivel upwards holding the persons or animals and preventing them from being rolled over by the car. This came along during an era when there was so much street railway traffic that the Cleveland Railway Company was able to advertise "a car in sight at all times."

Car 4049 was the only modern Cleveland streetcar to sport the famous Eclipse fender, called a cowcatcher by riders. *(Bill Vigrass photo)*

### End of an Era

CTS was most eager to say good-bye to the great era of the streetcar. Indeed, the parade signaling the demise of Euclid Avenue cars on April 27, 1952, was called a "celebration" and named the "Parade of Progress." Thereafter, as quickly as buses could be obtained, the remaining streetcar lines were converted. Lorain cars went in June 1952, East 55th and Superior in March 1953, and Clark and West 25th in August 1953. The supply of new buses was then exhausted, and so the one remaining car line, Madison Avenue, earned a reprieve until the next shipment of GM buses would arrive in January 1954.

On September 20, just one month after the Clark and West 25th lines had quit, a simmering smudgepot of sewer gas that had slowly

been gathering under West 117th Street exploded, ripping up the entire center of the street, including the entrances to the Madison Car House, and forcing suspension of streetcar service. CTS management immediately contacted suppliers in search of demonstrator buses that could be secured to keep the streetcars out of service, but only a few could be acquired.

Street repairs also put the Detroit Avenue trackless trolleys out of service, because their overhead wires did not permit them to negotiate around the repair areas. They had to be replaced by buses, and after hours Clifton Boulevard trackless trolleys were forced to park on the street because they could not get to Madison Station. Using buses on Madison and Detroit put a heavy strain on equipment availability.

CTS made a decision to cut service on some other lines to provide the buses needed on Madison and Detroit, but when *Cleveland News* reporter Harry Christiansen heard about the plan, he warned CTS general manager Donald Hyde that he would write a damaging story about how CTS was intentionally cutting service and causing hardships when streetcars were still available to take up the slack. Streetcars could only be used on Madison, as the tracks on Clifton and Detroit were long gone. Hyde relented and ordered track repair at the West 117th–Madison intersection.

With the westbound track repaired, CTS returned streetcar service to Madison Avenue at midnight, November 22, 1953, using eleven 4000-type cars, augmented by some buses. Adding to this hybrid operation was the fact that the streetcars had to run deadhead from Denison Station and operate single-track over the West 117th–Madison intersection using temporary crossovers, because tracks into Madison Station had been completely severed during repairs.

To carry out this complicated operating scheme, buses from other lines, demonstrator buses, and streetcars were all required to provide service during the last few weeks of rail operation. Eight additional streetcars were called out of retirement to round out the service requirements. These last nineteen cars were: 4051, 4057, 4062, 4082, 4094, 4105, 4108, 4109, 4110, 4115, 4117, 4119, 4131, 4135, 4136, 4138, 4142, 4144, 4145.

The part of the story that Hyde never knew, however, was the real reason Christiansen wanted the streetcars returned to Madison service. Christiansen, a consummate railfan, had just purchased a 35mm camera and wanted to complete a color slide coverage of the Madison line.

On January 23, 1954, car 4051 made the last revenue run on route 25, Madison, and the next day red-white-and-blue-bunted streetcars ran several special trips for a well-advertised "end-of-the-line" bash, from Public Square to West 65th Street and Madison Avenue. At 3:55 P.M. car 4142, operated by Donald Hyde, squealed around the south-

On March 24, 1953, the A. Shaw Co. scrapped the articulated cars en masse. Smoke from this funeral pyre could be seen for miles. Almost all Cleveland streetcars suffered the same fate. *(Roy Bruce photo, Blaine Hays Collection)*

west quadrant of Public Square for the last streetcar trip in Cleveland and in the State of Ohio.

Forty years after that dismal day in 1954, on September 16, 1994, in the Cleveland municipal parking lot, groundbreaking took place for the Regional Transit Authority's Waterfront Line. Built as an extension of the old Shaker Heights rapid transit trolley line, the new line brings light rail vehicles (read that as modern-day trolleys) back to a downtown operating loop. The squeals of steel wheels would be heard once more.

PART B

## Cleveland Rapid Transit Roster of Equipment

### SECTION I

#### Shaker Rapid Transit

#### Fleet Statistics: Center-Entrance Drop-Center Cars, 1–36

BLDR: G. C. Kuhlman Car Co. 1914–15; W: 50,000 lbs; LEN: 51'; seats: 60; 1200–1219 had low center conductor's platform and 1220–1232 had steps up to conductor's platform; CR 150–164 and 484–499 lost Brill 68 TRKS to SHRT cars in return for outslung 51-E TRKS; slot was left for 1200 to be REN 9 (with MU cables) instead of 0;

SHRT 30, FN 1230, REB with front door, G. C. Kuhlman, 1914. *(Blaine Hays Collection)*

paint schemes: "A" designates cars PTD in "Cleveland and Youngstown" paint, a green/yellow/gray with sunburst front design adopted by Van Sweringens to indicate the dawn of a new day and the lushness of the real estate they were selling along the boulevards; "B" designates the CIRR scheme of red/yellow/gray applied to the cars between 1936 and the end of World War II; "C" designates the gray/yellow scheme applied to the cars during municipal ownership. All cars were purchased from CR on 4-15-40. Pre-Shaker information on the fleet can be found in Part I of the roster. After 11-2-59 retirement, two trains made frequent appearances for special events and rush-hour service. One train consisting of cable-MU couplers was 1-2-6-11-12. The other, a button MU train consisted of 17-18-22-25-26 and spare 27. These were short-lived holiday extras and sporting event specials. CR T 22.

1200    A, (I), in CIRR service 1920, not MU, REN 1300 (II) 1922, returned to CR

1201    A, (I), in CIRR service 1920, not MU, REN 1299 (II) 1922, returned to CR

1202    A, (I), in CIRR service 1920, not MU, REN 1298 (II) 1922, returned to CR

1200    A, (II), MU cable, in SHRT service 1922, stored at Miles Station 1930, in service 11-1-42, used in Miles-Broadway to Bedford service 4-1-32 to 4-9-32, FN 1300, REN 9 on SHRT, B, OOS 1954, SCR 1955

1201    A, (II), MU cable, in CIRR service 1922, FN 1299, REN 1, B, C, OOS 1960, sold to Warehouse Point Museum 1960

1202    A, (II), MU cable, in CIRR service 1922, FN 1298, REN 2, B, C, OOS 1960, SCR 1960

1203    A, MU cable, REN 3, (originally non-MU, only car of original four cars to be CON) Standard Steel TRKS until 1940s, first car to operate on Moreland division 4-11-20, stored at Miles Station 1930, used in Miles-Broadway to Bedford service 4-1-32 to 4-9-32, in service on CIRR 10-2-40, B, C, stored 1957 after being flooded, motors removed after flood, became rubbish car at Van Aken Yards, sold to NORM 1968, moved to Trolleyville 1974, moved to NORM 1979; car now has 68-E1 TRKS

1204    A, MU cable, in CIRR service 1920, stored at Miles Station in 1930, used in Miles-Broadway to Bedford service 4-1-32 to 4-9-32, REN 4, B, C, in service on SHRT 6-15-41, OOS 1949, returned to service 1954, SCR 1960

1205    A, MU cable, in CIRR service 1920, REN 5, B, OOS 1954, SCR 1955

1206    A, MU cable, in CIRR service 1920, Standard Steel TRKS until 1940s, REN 6, used in Miles-Broadway to Bedford service 4-1-32 to 4-9-32, in SHRT service 6-15-41, B, C, OOS 1960, SCR 1960

1207    A, MU cable, in CIRR service 1920, REN 7, B, stored at Miles Station 1930, in CIRR service 7-15-42, OOS 1954, SCR 1955

1208    A, MU cable, in CIRR service 1920, Standard Steel TRKS until 1940s, REN 8, B, OOS 1954, SCR 1955

1209    A, MU button and MU cable, in CIRR service 1920, REN 36, B, OOS 1954, SCR 1956

1210    A, MU cable, in CIRR service 1920, Standard Steel TRKS until 1940s, REN 10, B but did not receive red belt rail stripe, became wrecker in 1949, painted light-blue and then light-green, flooded in 1957 and SCR

1211    A, MU cable, in CIRR service 1920, REN 11, B, C, became replacement wrecker to 10 probably 1960, received pantograph in RTA days, sold to NORM 1981

1212    A, MU cable, in CIRR service 1920, REN 12, B, C, B, OOS

1960, reserved for excursions in May 1963 and PTD by Euclid Railfans' Club into B red/gray/yellow scheme; only intact streetcar to survive into RTA era; equipped with pantograph; REN on paper in 1978 as "012 Tow Car" and continues in that capacity to this day; last car into CUT Traction Concourse 12:13 AM 12-10-90, operator Tim O'Donnell

1213   A, MU cable, in CIRR service 1920, REN 13, B, C, OOS 1954, SCR 1956

1214   A, MU cable, in CIRR service 1920, Standard Steel TRKS until 1940, REN 14, B, OOS 1949, SCR 1954

1215   A, MU button, in CIRR service 1923, REN 15, B, first car into CUT Traction Concourse 6:37 am, 7-20-30, became storage shed at Kingsbury 1954, PTD white, SCR 1962

1216   A, MU button, in CIRR service 1923, REN 16, B, OOS 1955, SCR 1956

1217   A, MU button, in CIRR service 1923, REN 17, B, C, sold to CTS in 4-1960 and CON to line car 024, car still active on RTA

1218   A, MU button, in CIRR service 1923, serious front-end ACC 1929, REN 18, B, C, OOS 1960, sold to Gerald E. Brookins Museum of Electric Railways 7-1961, moved to Trolleyville where it is in storage

1219   A, MU button, in CIRR service 1923, REN 19, B, OOS 1954, SCR 1954

1220   B, MU button, in CIRR service 1925, REN 20, OOS 1954, SCR 1954

1221   B, MU button, in CIRR service 1925, REN 21, OOS 1954, SCR 1955

1222   B, MU button, in CIRR service 1925, REN 22, C, OOS 1960, SCR 1960

1223   B, MU button, in CIRR service 1925, REN 23, OOS 1953, SCR 1954

1224   B, MU button, in CIRR service 1926, REN 24, OOS 1955, SCR 1955

1225   B, MU button, in CIRR service 1926, REN 25, C, stored at Miles Station in 1930, in CIRR service 12-15-40, sold to Gerald E. Brookins Museum of Electric Railways 7-1961, moved to Trolleyville where it continues in service

1226   B, MU button, in CIRR service 1926, REN 26, C, OOS 1960, control group and other parts sold to CTS 4-1960 to make second end of line car 024, remainder SCR 1960

1227   B, MU button, in CIRR service 1926, possibly used in test service over western lines of Southwestern interurban in 1926 or 1927, REN 27, stored at Miles Station 1930, in CIRR service 1936, C, was part of "bankers' special" train (consisting of 27, 28, 30, 32) and received leather seats from Kentucky

Traction and Terminal (KT&T) car, OOS 1960; originally ACQ by National Capitol Trolley Museum, given by them to Jim Lilly of Trolley Valhalla of Tansboro, NJ, and shipped to a storage location in Baltimore, MD, 1961; car frame badly split in transport during move, shipped to Tansboro site in 1968 where car deteriorated to very poor condition; car ACQ by Ron Jedlicka of Buckeye Lake Trolley of Thornville, OH, and given to Seashore Trolley Museum without TRKS on condition it would be restored; car now restored at Seashore Trolley Museum, Kennebunkport, ME

1228    B, MU button, in CIRR service 1927, KT&T seats, REN 28, REB with front door as OM 1949, front coupler removed, never used, stored Van Aken loop 1949, SCR 1955

1229    B, MU button, in CIRR service 1927, REN 29, stored at Miles Station in 1930, in CIRR service 1936, OOS 1954, SCR 1955

1230    B, MU button, in CIRR service 1927, KT&T seats, REN 30, first car REB with front door OM 9-1945, deadman CTL added and front coupler removed, OOS 1955, SCR 1955

1231    B, MU button, in CIRR service 1927, REN 31, REB with front door as OM 1949 and deadman CTL added, C, OOS 1955, CON to wire greaser by construction of flat roof and addition of grease trolley pole 1955, CON to wrecker by RTA, pantograph and RTA paint added 1977, body only sold 1992 to Brian Maher, ACQ by Bill Dunkel and moved to Medina Hobby Shop on RT 18 where it sits atop freight car TRKS; PTD green in 1994, ACQ NORM 3-16-96

1232    B, MU button, in CIRR service 1927, KT&T seats, REN 32, REB with front door as OM 9-1945 and deadman CTL added, OOS 1954, SCR 1955

### Fleet Statistics: Shaker Rapid Transit Trailers

1233, 1234, 1235 • BLT by G. C. Kuhlman Car Co.; and 2318, 2319, 2365 BLT by CR.

1233    B, MU button, in CIRR service 1927, stored at Miles Station in 1930, in SHRT service 11-1-42, REN 53, OOS 1948, REN 46 when CON to shop shed, PTD green then brown, SCR 1974

1234    B, MU button, in CIRR service 1927, stored at Miles Station 1930, in SHRT service 3-15-43, REN 54, OOS 1948, REN 47 when CON to shop shed, PTD green then brown, SCR 1974

1235    B, MU button, in CIRR service 1927, stored at Miles Station in 1930, towed to Fairmount Station 1935 and used as office during reconstruction, then returned to Miles, car OOS for 17 years! in SHRT service 1947, REN 55, OOS 1948, REN 45, PTD green when CON to shop shed for sand and salt storage, REN 43, PTD brown, SCR 8-1976

2318   B, MU cable, in SHRT service 1947, REN 56, REN 44 when CON to shop shed, given to Ron Jedlicka of Buckeye Lake Trolley in Thornville, OH, moved to Homerville, OH, donated to Seashore Trolley Museum on condition it be restored as mate to motor 27, car in storage at Seashore Trolley Museum, Kennebunkport, ME

2319   B, MU cable, in SHRT service 1947, REN 57 in 1947 but never used in service, REN 45 (II) when CON to storage shed at shop, OOS 1948, SCR 1974 at Kingsbury Shops

2365   B, MU cable, in SHRT service 1947, REN 58, C, OOS 1948, towed to Van Aken Yards and made permanent trainmen's locker room and office, towed to Kingsbury Shop 4-3-68, sold to Gerald E. Brookins Museum of Electric Railways and moved there 1968, PTD CR yellow, car 80% restored to CR appearance including Peter Smith coal stove, car in storage at Trolleyville

*Note:* Series numbers 33–35 were originally left vacant in renumbering in the event 53–55 were ever remotorized.

### Fleet Statistics: Shaker Rapid Peter Witts and Other Loaned Cars

*Units marked with an asterisk were not numbered.*

4050–4051   A, CR cars equipped with air whistle and painted in variation of Van Sweringen green "sunburst" paint in 1935 for OM test runs on CIRR during period when CR was under Van Sweringen control

5014   articulated CR car equipped with pantograph in 1930 and made trial run on CIRR during period CR was under Van Sweringen control, car could not clear loop under Union Terminal and was returned to CR

★   a Northern Ohio Traction and Light Company steel interurban car made a run on the Moreland line in 1928 to Public Square to test clearances in the event the Van Sweringen plans for bringing the interurbans into downtown ever materialized

409   Red Arrow Railbus, GM "fishbowl" bus 409 made test run over Shaker lines in 1967 using rail bogies manufactured by Fairmount Equipment Co. weighing 5,000 lbs.

3201   Boston Pullman PCC car operated in test service over the Shaker line from 6-17-46 to 6-27-46

3401   Boston Boeing United States LRV operated in test service over the Shaker line from 7-29-76 to 8-5-76; upon arrival and acceptance in Boston, car ran only two and one-half months before a major derailment caused irreparable damage, car SCR later and outlived in service by PCC 3201

- ★ Leyland Railbus (40-seat, dual cabs, provided by British Rail-Leyland through Associated Rail Technologies) operated a scheduled demonstration service over RTA and Norfolk and Western routes between CUT (Shaker Rapid platform) and a N&W siding at Maple Street in Mentor for five weeks in April and May 1985 (24 miles)
- ★ Sperry Rail Service geometry car operated over entire RTA rapid system in 1981 to locate and diagram track defects

### Fleet Statistics: "Smokers," 300–306

BLDR: St. Louis Car Co. 1923, for Aurora, Elgin and Chicago RR Co.; W: 36,000 lbs; LEN: 38'; seats: 52; MOT: 4GE265; CTL: K35G; TRKS: BLT with St. Louis arch bar, then St. Louis 64-E-1-B; ACQ from Fox River upon 3-31-35 abandonment; partitioned smoking section. REB for CIRR at Harvard Shops by CR 1936. Under agreement with CTS, these cars last operated March 14, 1955, because trip arms were too difficult to install and they were slow and would have bogged down operations in the joint E. 55th to CUT CTS-SHRT trackage.

| | |
|---|---|
| 300 | in CIRR service 9-1936, OOS 1949, sold to Kenosha Motor Coach Lines (Milwaukee Speedrail) 9-1949 |
| 301 | B, in CIRR service 9-1936, OOS 1949, sold to Kenosha Motor Coach Lines (Milwaukee Speedrail) 9-1949 |
| 302 | B, in CIRR service 9-1936, OOS 1953, sold to Gerald E. Brookins Museum of Electric Railways 1954 for parts, SCR 1985 and salvaged for parts |
| 303 | B, in CIRR service 9-1936, OOS 3-14-55, sold to Gerald E. Brookins Museum of Electric Railways 1955, operates regularly there |
| 304 | B, in CIRR service 9-1936, OOS 1954, sold to Gerald E. Brookins Museum of Electric Railways 1955, operates regularly there |
| 305 | B, in CIRR service 9-1936, OOS 1950, burned-out MOT during big Thanksgiving snow, SCR 1954 |
| 306 | B, in CIRR service 9-1936, OOS 1953, sold to Gerald E. Brookins Museum of Electric Railways 1954 for parts and re-sold 1992 to Illinois Railway Museum where car was made operational |

## Fleet Statistics: Massillon Lightweights, 60–65

BLDR: Cincinnati Car Co.; "Curve Siders" had actual curvature to the side of the body; BLT for the Indianapolis and Southeastern interurban 1928–29; W: 38,000 lbs; LEN: 46' 1"; seats: 38; MOT: 4GE247; CTL: K75E7. Sold in 1932 to Intercity Rapid Transit Co. in Canton and used as Canton-Massillon parlor cars. Cars had a partitioned smoking section. Picked up by CIRR at a low price upon abandonment of Intercity on 12-17-40 and sold to the Kenosha Motor Coach Lines (Milwaukee Speedrail) in 9-1949.

SHRT 65, former Intercity Rapid Transit car, with smoking section, Cincinnati Car, 1929. *(Anthony F. Krisak photo, Northern Ohio Railway Museum)*

| | |
|---|---|
| 60 | B, unloaded Warrensville yards 4-10-41, FN 260, in CIRR service 11-16-41 |
| 61 | B, unloaded Warrensville yards 4-11-41, FN 235, in CIRR service 11-16-41 |
| 62 | B, unloaded Warrensville yards 4-11-41, FN 245, in CIRR service 11-16-41 |
| 63 | B, unloaded Warrensville yards 4-16-41, FN 200, in CIRR service 2-7-42 |
| 64 | B, unloaded Warrensville yards 4-15-41, FN 240, in CIRR service 3-25-42 |
| 65 | B, unloaded Warrensville yards 4-14-41, FN 230, in CIRR service 6-4-42, painted in gray-yellow "C" scheme as test and later adopted as standard colors for short fleet |

## Fleet Statistics: St. Louis Public Service PCC Cars, 40–49

BLDR: St. Louis Car Co. 1946; W: 36,420 lbs; LEN: 46 ft; seats: 55; MOT: GE1220; CTL: GE; St. Louis, MO, city streetcars in the 1766–1775 series. Cars REB to MU with couplers installed and PTD in Shaker yellow/gray/green by St. Louis Public Service Co. "Riot screens" were installed over windows in 1967 to prevent rocks from penetrating; trail stop lights and front dash lights used on cars in St. Louis were removed in 1960s. All cars OOS 4-6-82, except as noted, and moved to Kingsbury Yards 10-1984 for salvage.

*Note:* In 1974 SHRT began to repaint the PCC fleets and vary its traditional yellow/gray/green stripe design. Some cars were repainted in yellow with an orange stripe (YO), some were painted in bright traction orange (O), and two variations of RTA red/white/orange were also applied (R1 and R2). Affected cars are so designated in the roster.

SHRT 49, former St. Louis Public Service 1775, St. Louis Car, 1946. *(Alex Thiel photo, Blaine Hays Collection)*

40    FN 1771, shipped from St. Louis 5-13-59, arrived SHRT shop 5-18-59, in service 6-26-59, REN 50 in early 1975 in order to provide consecutive numbering of PCC cars 41–95, YO, OOS 4-1982, stored Westpark, sold Ron Jedlicka 1985, moved to Thornville, OH

41    FN 1769, shipped from St. Louis 5-13-59, arrived SHRT shop 5-18-59, in service 6-17-59, OOS 1-1982, sold to Ron Jedlicka 1983, moved to Thornville, OH 1984

42    FN 1767, shipped from St. Louis 7-6-59, arrived SHRT shop 7-13-59, in service 8-11-59, YO, stored Westpark 1982, TRKS and control equipment purchased by New Jersey Transit 1985, dumped on side and stripped 1988 and SCR at Kingsbury

43    FN 1774, shipped from St. Louis 7-6-59, arrived SHRT shop 7-13-59, OOS 6-1982, stored Westpark, sold New Jersey Transit in 1985, dumped on side and stripped 1988 and SCR Kingsbury

44    FN 1768, shipped from St. Louis 8-11-59, arrived SHRT shop 8-18-59, in service 9-22-59, OOS 4-1982, stored Westpark, sold to New Jersey Transit 1985, dumped on side and stripped, SCR 1988

45    FN 1772, shipped from St. Louis 8-11-59, arrived SHRT shop 8-18-59, in service 9-4-59, OOS 6-1982, stored Westpark, sold New Jersey Transit 1985, dumped on side and stripped, SCR 1988

46    FN 1770, shipped from St. Louis 9-22-59, arrived SHRT shop 9-30-59, ACC in Van Aken yard one week before line shut for renovation on 3-28-81, stripped for sale as cottage but no bids, SCR 1988 at Kingsbury

47    FN 1773, shipped from St. Louis 9-22-59, arrived SHRT shop 9-30-59, YO, stored Westpark, sold 5-1987 to National Capital Trolley Museum, traded to individual in Ozarks (AK) in return for body of DC Transit 1540

48    FN 1766, shipped from St. Louis 11-2-59, arrived SHRT shop 11-9-59, stored Westpark 1982, body burned by arsonists in 1985 at Kingsbury, sold 1987 Kingsway Construction, SCR 1988

49    FN 1775, shipped from St. Louis 11-2-59, arrived SHRT shop 11-9-59 (last car ACQ by SHRT), YO, ACC damage and OOS 10-1980, stored Westpark 1982, sold to National Capitol Trolley Museum, then traded to NORM and moved there 7-1987

### Fleet Statistics: Minneapolis–St. Paul PCC Cars 51–70, 3, and 27

BLDR: St. Louis Car Co. 1947, for Twin City Rapid Transit Co.; W: 37,990 lbs; LEN: 46' 3"; seats: 54; MOT: GE1220; CTL: GE. All cars ACQ in 1953 except 3 and 27, which were ACQ from New Jersey Transit with insurance money after the wreck of 59 and 65 in 1977. All cars OOS by 12-1982, except 54, 70, and others that operated on a limited basis during Christmas holidays in 1984 and 1985. All cars MU except where noted. All cars equipped with pantographs placed over the lead truck, except 3 and 27. While in city service, cars had tail lights that were covered over and in late 1970s were reinstalled. Non-MU cars 51–55 known as "doodle-bugs" on Shaker Rapid.

51  FN 340, non-MU, in SHRT service 7-12-54, first car PTD traction orange 12-20-74, R2, sold 1991 Niagara Frontier Transit Authority (NFTA), Buffalo, and stored

52  FN 341, non-MU, in SHRT service 9-9-54, YO, O, R2, had dark-green under front anti-climber, sold 1991 NFTA and stored

53  FN 342, non-MU, in SHRT service 9-6-54, PTD blue/gold in 1962 to mark fiftieth anniversary of City of Shaker Heights, YO, O, R2, sold 1991 NFTA and stored

54  FN 343, non-MU, in SHRT service 8-2-54, O, R2, sold 1991 NFTA and stored

55  FN 344, non-MU, in SHRT service 3-31-54, YO, O, R2, sold 1991 NFTA and stored

56  FN 345, in SHRT service 4-30-54, O, R2, last car PTD in RTA colors 1981, sold 9-1982 Vintage Trolley Restoration, MD

57  FN 346, in SHRT service 4-26-54, damaged in shop collision, YO, O, R2, OOS 1-1982, stored CUT, sold Kingsway Construction, SCR 1987 with TRKS to Seashore Trolley Museum

58  FN 347, in SHRT service 5-4-54, YO, O, R2, OOS 5-1982, sold 9-1982 Vintage Trolley Restoration, MD

59  FN 348, in SHRT service 4-23-54, O, destroyed in head-on collision 7-8-77, sold Kingsway Construction and SCR 1987

60  FN 349, in SHRT service 4-24-54, O, R2, sold 1991 NFTA and stored

61  FN 350, in SHRT service 6-3-54, YO, O, R2, sold 1991 NFTA and stored

62  FN 351, in SHRT service 4-23-54, O, R2, OOS 5-1982, sold 1991 NFTA and stored

63  FN 352, in SHRT service 6-14-54, YO, O, long-term lease to Trolleyville, USA, where it is operational

64  FN 353, in SHRT service 5-4-54, YO, O, R2, sold 9-1992 Ohio Railway Museum for parts

65  FN 354, in SHRT service 6-21-54, O, destroyed in head-on collision 7-8-77, sold Kingsway Construction and SCR 1987

66  FN 355, in SHRT service 6-11-54, O, R2, OOS 5-1982 after electrical fire, held by RTA pending lease to NORM

67  FN 356, in SHRT service 4-30-54, o, R2, sold 1991 NFTA and stored

68  FN 357, in SHRT service 6-19-54, O, R2, last car to run in traction orange paint, and last car to operate over Van Aken line before REB 3-28-81, sold 1991 NFTA and stored

| | |
|---|---|
| 69 | FN 359, in SHRT service 6-17-54, O, sold 1991 NFTA and stored |
| 70 | FN 360, in SHRT service 6-14-54, YO, O, R2, "Care Bears" Christmas car 1984–85; at suggestion of Morris Stone of American Greetings Corp., PTD white and decorated with wreath around headlight; last PCC car to run in revenue service on Shaker 12-21-85; sold 1991 NFTA, used there in test runs with pantograph, then stored |
| 3 | FN 322, arrived from Newark, NJ 11-16-77, operated in red-white-blue Newark scheme until PTD R2; transferred to Minnesota Transportation Museum 8-1990 |
| 27 | FN 416, arrived from Newark, NJ 11-17-77, same paint as car 3, R2, transferred to Minnesota Transportation Museum 8-1990 |

### Fleet Statistics: Pullman PCC Cars, 71–95

BLDR: Pullman-Standard 1948, last order of cars to be built at Worcester, MA. All cars D between 10 and 11-1948. W: 43,100 lbs; LEN: 50'; seats: 58; MOT: GE1220; CTL: GE. Cars BLT with a door in the left-center for use in Cleveland's downtown subway then under serious consideration; door never used and sealed off with bench seat to cover it. Fleet's dash lights removed in 1960s. All cars were delivered in yellow paint with three green stripes, at the request of Shaker Heights mayor William Van Aken, who got the idea from Cincinnati PCC cars. Later, when cars were repainted, there were variations with striping. The following cars had but one green stripe: 71, 78, 81, 83, 86, 89; all others were given two green stripes. Cars BLT with full skirting, but during RTA days skirting was cut around TRK area (except on 76) to ease wheel grinding and provide maintenance accessibility. All cars OOS 6-9-82. This fleet purchased under an equipment trust; all cars carried the following stencil: "City of Shaker Heights, equipment trust of 1948, Central National Bank of Cleveland, trustee, owner and lessor."

| | |
|---|---|
| 71 | could not operate MU after 1971 due to damaged coupler, stored CUT, sold to Trolleyville, USA |
| 72 | stored CUT and sold to NORM |
| 73 | stored CUT, sold to Ed Metka 1987 and moved to Trolleyville, USA, moved to Hagerstown, MD 1988 |
| 74 | carried banners proclaiming "Rapid RTA 35¢" in 1975, REB 1978 eliminating left side door, YO, R1, sold to Trolleyville, USA 1985, SCR 1996 |
| 75 | carried banners proclaiming "Rapid RTA 35¢" in 1975, REB 1978 eliminating left side door, YO, R1, sold to Trolleyville, USA 1985 |

SHRT 71, Pullman-Standard 1948. *(Blaine Hays photo)*

76     PTD white with tiny red stripes on the belt rail and a gray roof at request of general manager Leonard Ronis in 1976, left side door removed during renovation and large window installed; flower box installed under window with green sign proclaiming "RTA WELCOMES SPRING" at request of special projects chief Harry Christiansen; car retained for historical fleet, OOS 1-1982, stored Westpark 6-1982, dead storage E. 55th yards 1996

77     stored CUT, sold to Ed Metka 1987 and moved to Trolleyville, USA 1988, moved to Hagerstown, MD

78     stored Westpark, sold to NORM 1987

79     could not operate MU after 1971 due to coupler damage, stored CUT, sold Ron Jedlicka 1985 and moved to Thornville, OH

80     skirts cut, given yellow deluxe Imron paint, YO, stored Kingsbury 6-1982, sold to Ron Jedlicka 1985, moved to Thornville, OH

81     last car to operate on Shaker Blvd. line before REB 4-19-80, stored Kingsbury, sold to Seashore Trolley Museum for salvage of electrical gear but moved to NORM 1987, parts later exchanged with Seashore Trolley Museum

82     carried banner proclaiming "Rapid RTA 35¢" in 1975, YO, stored Kingsbury, sold to East Lansing Railway Museum, exchanged with NORM for parts salvage, SCR Brookpark 1988

83     stored Kingsbury, sold Seashore Trolley Museum for salvage of electrical gear, car moved to NORM and parts from it traded to acquire SHRT PCC 49

84     carried banner proclaiming "Rapid RTA 35¢" in 1975, experimental paint applied in 1978 and window replaced in left side door, roof trim shaved upward giving car the name

| | |
|---|---|
| | "Mohawk," R1, stored at Westpark, sold to Ron Jedlicka 1985 and moved to Thornville, OH |
| 85 | carried banner proclaiming "Rapid RTA 35¢" in 1975, YO, ACC 1978 rear-end damaged and repaired, stored Kingsbury, sold Ed Metka 1987 and moved to Trolleyville, USA, as a gift for allowing storage of other cars |
| 86 | stored Westpark, sold to NORM 1985 |
| 87 | ACC 12-1966 (see CTS roster, cars 255-256), stored Kingsbury, sold to Ed Metka 1987 and moved to Trolleyville, USA, moved to Hagerstown, MD 1988 |
| 88 | OOS 3-1981 due to major CTL problem, stored Kingsbury, sold 1983 for resale as cottage, stored E. 55th yard but no bids, SCR 1983 |
| 89 | stored Kingsbury, sold to Ed Metka 1987 and moved to Trolleyville, USA and then to Hagerstown, MD 1988 |
| 90 | YO, R1, carried banners proclaiming "Rapid RTA 35¢" in 1975, used as emergency passenger shuttle 7-8-77, 8-1977 involved as runaway car downhill from Shaker Square, stored Westpark, sold to Ron Jedlicka 1987, SCR for parts at Brookpark yards 1-1989 |
| 91 | stored Kingsbury, sold Ed Metka 1987 and moved to Trolleyville, USA, then to Hagerstown, MD 1988 |
| 92 | REB and picture window added, stored Kingsbury, sold to NORM 1987 |
| 93 | stored Westpark, sold to Ed Metka 1987 and moved to Trolleyville, USA, and then to Hagerstown, MD 1988 |
| 94 | stored Kingsbury, sold to Pennsylvania Trolley Museum and slated for REB for operating left side door |
| 95 | R1, REB and picture window added, OOS fire damage 1976, stored Westpark, sold Ron Jedlicka 1985 and moved to Thornville, OH |

### Fleet Statistics: Illinois Terminal Double-Ended PCC Cars 450–451

BLDR: St. Louis Car Co. 1948; W: 42,680 lbs; LEN: 50' 5"; seats: 55; MOT: GE1220; CTL: GE. Operated on RTA as a two-car train at times.

| | |
|---|---|
| 450 | Illinois Terminal DE PCC car, leased from Ohio Railway Museum of Worthington and arrived on SHRT 9-1975 for possible direct airport service and even carried "airport" destination sign, REB electrical and body on Shaker, in service on SHRT 1-1978, returned to museum on flatcar HTTX 91222 on 5-8-79, ACC upon return to ORM, repaired 1992 |
| 451 | Illinois Terminal DE PCC car, leased from Warehouse Point, CT and arrived on SHRT 9-1975 for MU use with 450 for |

airport service, REB electrical and body on Shaker, in SHRT service 1-1978, returned to museum by flatbed truck 5-3-79

### Fleet Statistics: CTS-TTC PCC Cars

BLDR: Pullman-Standard 1946; LEN: 46' 5"; W: 39,800 lbs; seats: 53; MOT: 4WH1432J; CTL: WH; cost: $17,500 each. (This might have been $9,000 each.) Cars purchased from Toronto Transit Commission in 1978 for winter service; all removed from service 3-28-81, final day of operation of Van Aken line before rebuilding.

4630     to Cleveland RTA 9-6-78, never used, sold to Ohio Railway Museum 1982, SCR for parts for Cincinnati PCC ACQ from Toronto

4648     to Cleveland RTA 8-15-78, first to operate on RTA in "Welcome Home" run along with 4652 signifying its return to Cleveland rails, sold to Illinois Railway Museum 1982 where it is undergoing restoration

4651     to Cleveland RTA 9-8-78, sold to Illinois Railway Museum 1982, SCR for parts

4652     to Cleveland RTA 8-8-78, operated in "Welcome Home" run, ACC 4-1980 not repaired, sold to NORM 1982, SCR Kingsbury 1983 for parts

4655     last PCC to leave CTS in 1953 and returned to Cleveland 9-19-78, sold to NORM 1982, slated for restoration to CTS colors and REN 4230

4656     to Cleveland RTA 10-25-78, sold to NORM 1982, moved to museum 9-11-83, slated to be restored as Toronto car

4662     to Cleveland RTA 11-3-78, last to run on RTA 1980, ACQ by Ron Jedlicka 1982 and moved to Thornville, OH

4663     to Cleveland RTA 11-17-78, ACQ by Ron Jedlicka 1982 and moved to Thornville, OH and SCR

4665     to Cleveland RTA 12-19-78, ACQ by Mad River and NKP Railroad 1982, moved to Bellevue, OH 1983

### Fleet Statistics: Breda Articulated Cars, 801–848

BLDR: Breda Costruzioni Ferroviarie, Pistoia, Italy; LEN: 77'; seats: 84; W: 84,000 lbs; MOT: Brown Boveri 468 kilowatt; CTL: Brown Boveri; AC; cost: $625,000 each; D began 10-1979, all accepted between 8-1981 and 11-1982. Cars 801 to 805 designated Waterfront Cars in 1996 and PTD a blue/turquoise livery and decorated for the city's bicentennial celebration; cars 804, 813, 823, 830, 836, 847 fitted as ice trains with special ice-cutting pantographs for severe winter weather; cars 819, 822, 824, 829, 838, 843, 846 still in original paint in 1996.

801     (mock) one-third car body for design and acceptance purposes

802, 810, 812, 813, 814 • in-service inaugural day 10-31-81; 802 first car D 11-26-80 due to 801 being used in test service in Italy; 810 first car in revenue service on the Green line

804     AC test, car shipped to United Fruit Growers plant in Richmond, VA, for cold-air testing

805     ACC 12-10-83

816     ACC, sideswiped by 813 on E. 75th Street 10-11-93, shipped to Delaware Car Co. for REB

818     ACC collision with 840 in E. 55th yard 9-1984, "A" end stored, "B" end became 849, repairs to be made in-house

838     ACC 12-1982

840     ACC with 818 at E. 55th yard 9-1984, "B" end stored, "A" end became 849, "B" became part of 816, and REN 816

849     combination of 840A and 818B entered service 12-1984, REN 840 when repairs to 816 and 818 completed

*Notes:* Shortly after D of first cars, Cleveland was descended upon by a Philadelphia-style graffiti invader named "The Ranger," who covered the sides of several of the new cars plus some of the Pullman Air-porter fleet with colorful writing. Fortunately this trend was short lived, but the phantom transit painter was never apprehended. In 1995–96, to prepare the fleet for service on the new Waterfront line, the cars were equipped with high-reach pantographs for Conrail crossings and provided with wheelchair access to meet requirements of the Americans with Disabilities Act. Cars 801–805 were particularly designated as "Waterfront cars" and dressed up in special paint. Some cars in this fleet were involved in a sale-leaseback agreement whereby equity interest was purchased by private firms for income tax purposes, resulting in RTA receiving money from firms and the firms getting a tax break.

SECTION 2

*Cleveland Rapid Transit*

*Fleet Statistics: Kentucky Traction and Terminal Cars*

BLDR: Cincinnati Car Co. 1922–23, curve-side design, purchased used from KT&T by the Van Sweringen brothers in 1934 for use on

their East Cleveland Line, otherwise known as Nickel Plate East; TRKS: Cincinnati Arch Bar; W: 25,180 lbs; seats: 44; LEN: 40'3". Line was not completed and cars not used. Cars moved from Harvard Shops to Kingsbury Shops in 1937 when Van Sweringen interests lost control of CR. A test run indicated that the cars did not have the power to climb the grade between Woodhill Road and Shaker Square. Parts from the cars were used to upgrade the Shaker 60-series Canton-Massillon fleet. Cars were numbered 300–308 and 310. SCR 1941. One TRK survived into 1980s for shop use.

300, purchased by Van Sweringens from Kentucky Traction and Terminal for their East Cleveland line, Cincinnati Car, 1923. *(Eugene Schmidt Collection, Northern Ohio Railway Museum)*

### *Fleet Statistics: "Bluebird" PCC Rapid Transit Cars, 101–118, 201–270*

BLDR: St. Louis Car Co., 101–112, 201–256 in 1954; 113–118, 257–270 in 1958. Cars 101–118 CTS F 02; cars 201–270 CTS F 05. TRKS:

General Steel Industries; MOT: 4WH 55hp; CTL: Westinghouse Cineston (controlled acceleration, coasting, braking); LEN: single 48' 6", double 97' 6"; W: 56,000 lbs; seats: 54; cost: (1954) $67,400 and $117,700, (1958) $83,120 and $152,820. When ordered, cars numbered 1–56 (REN 201–256) and 57–68 (REN 101–112) with first numbers on paper only. Car order was originally to be for 58 cars but was reduced to 56 in 1953. Cars 101, 105, 108, 109, 113 retained by RTA. Known as PCC cars because they had Transit Research Corp. patented TRKS that were designed as a refinement of the Clark-built TRKS made for PCC streetcars. "Bluebird" name resulted from employees calling them that in later years, but they were always known to railfans as "Spam Cans." Cars PTD CTS dark-blue/silver/gray/red and starting in 1967 and concluding in 1968, certain cars (noted with ★) were PTD a lighter CTS blue. All cars D to Cleveland from St. Louis via Nickel Plate Railroad.

| | |
|---|---|
| ★101 | loaded 2-8-55 on flat PM16811, D SHRT E. 49th 2-17-55, PTD aluminum band 5-6-59, PTD 7-29-65, OOS 12-1984, stored Brookpark |
| ★102 | loaded 2-14-55 on flat PM16821, D SHRT E. 49th 2-23-55, PTD 8-13-65, derailed 6-7-55, air compressor fell out in service causing considerable body damage, OOS 9-1977, transferred Kingsbury, sold for SCR 1979 |
| ★103 | loaded 2-22-55 on flat ATSF91613, D SHRT E. 49th 3-2-55, PTD aluminum band 10-16-58, PTD 8-26-65, OOS 12-1984, stored Brookpark, SCR 1988 |
| ★104 | loaded 2-22-55 on flat NYC500700, D SHRT E. 49th 3-1-55, PTD aluminum band 5-6-59, had pantograph electrical control problem, electrical part burned hole through roof, never repaired properly (sat at Windermere shop for so long employees adorned it with funeral wreath), SCR 1980 |
| ★105 | loaded 2-28-55 on flat NYC500454, D SHRT E. 49th 3-8-55, PTD aluminum band 4-28-59, OOS 12-1984, partially ReB and stored Brookpark |
| 106 | loaded 2-28-55 on flat car PRR473894, D SHRT E. 49th 3-8-55, PTD aluminum band 5-1-59, ACC 1970, car end pushed in, SCR 1970 |
| ★107 | loaded 3-4-55 on flat car PM16793, D SHRT E. 49th 3-10-55, PTD aluminum band 4-23-59, ptd 9-2-65, OOS 1982, end pushed in at airport station against bumper post, stored Windermere, SCR 1984 |
| ★108 | loaded 3-4-55 on flat car NYC500327, D SHRT E. 49th 3-24-55, PTD 10-2-58, OOS 12-1984, stored Brookpark |
| ★109 | loadE.d 3-9-55 on flat car PRR475216, D SHRT E. 49th 3-14-55, PTD aluminum band 4-16-59, OOS 12-1984 |

CTS Bluebird 268, St. Louis Car, 1958. *(Blaine Hays photo)*

             stored Windermere, moved to Central Rail shop for renovation into garbage train 2-1990, PTD RTA colors white/orange/red, stored inside Brookpark shop (see service cars)

★110      loaded 3-9-55 on flat car NYC500507, D SHRT E. 49th 3-14-55, PTD aluminum band 4-28-59, damaged by fire 1976, sold for SCR 1979

★111      loaded 3-18-55 on flat car NYC499664, D SHRT E. 49th 3-24-55, PTD aluminum band 4-23-59, SCR 1979

★112      loaded 3-18-55 on flat car NYC500243, D SHRT E. 49th 3-24-55, PTD aluminum band 4-16-59, PTD 7-7-65, OOS 12-1984 stored Brookpark, to NORM 1988

★113      loaded 1-9-58 on flat car B&O8941, D SHRT E. 49th 1-14-58, PTD 6-8-65, OOS 12-1984, stored Brookpark

★114      loaded 1-17-58 on flat car WAB189, D SHRT E. 49th 1-27-58, PTD 6-16-65, severely damaged by fire 9-9-69,

| | |
|---|---|
| | repaired 1-1970, OOS 1982 stored Brookpark, transferred to Kingsbury 1985, SCR 1987 |
| ★115 | loaded 1-24-58 on flat car NYC499771, D SHRT E. 49th 1-30-58, PTD 6-29-65, OOS Kingsbury and sold for SCR 1979 |
| 116 | loaded 1-29-58 on flat car ERIE8126, D SHRT E. 49th 2-4-58, PTD 6-8-65, CTL problems 1976, OOS 12-1984 stored Windermere, SCR 1984 |
| 117 | loaded 1-31-58 on flat car B&O8974, D SHRT E. 49th 2-6-58, PTD 6-25-65, OOS ACC 12-1984 stored Windermere, SCR 1984 |
| 118 | loaded 2-7-58 on flat car SP143528, D SHRT E. 49th 2-13-58, PTD aluminum band 10-16-64, OOS 12-1984 with bad wiring, stored Windermere, SCR 1984 |
| 201–202 | loaded 9-21-54 on flat cars NYC499898 and NYC-500714, D SHRT E. 49th 9-27-54, car 202 used in testing for derailment practice 5-26-55 (intentionally put on ground and rerailed), PTD aluminum band 5-8-59, OOS at Windermere 1985, sold to NORM 1988 |
| ★203–204 | loaded 9-24-54 on flat cars NYC 500711 and ACL77061, D SHRT 10-1-54, PTD aluminum band 5-21-59, ACC derailed in Windermere yard 5-31-55, PTD 9-12-66, OOS 1984 stored Kingsbury, SCR 1987 |
| ★205–206 | loaded 9-29-54 on flat cars SP143011 and EJE6492, D SHRT E. 49th 10-6-54, PTD aluminum band 5-13-59, OOS 1-1985 due to fire, stored Windermere, sold to Indiana Railroad at Indianapolis, shipped by rail 6-10-88 |
| ★207–208 | loaded 10-1-54 on flat cars NYC499412 and CGW2133, D SHRT E. 49th 10-8-54, PTD aluminum band 5-18-59, OOS 3-1985, stored Windermere, SCR 1987 |
| ★209–210 | loaded 10-5-54 on flat cars NYC499878 and PM16677, D SHRT E. 49th 10-13-54, ACC joint area damage to 210 5-22-57, PTD aluminum band 5-26-59, PTD 7-1-66, OOS 1988 stored Brookpark, SCR 5-1988 |
| 211–212 | loaded 10-8-54 on flat cars PM16777 and CGW2064, D SHRT E. 49th 10-14-54, PTD aluminum band 5-29-59, OOS 1987 stored Brookpark, SCR 1988 |
| ★213–214 | loaded 10-14-54 on flat cars NYC500166 and RI90820, D SHRT E. 49th 10-20-54, PTD aluminum band 6-3-59, PTD 7-6-67, OOS 1987 stored Brookpark, SCR 1988 |
| 215–216 | loaded 10-29-54 on flat cars NYC499875 and KCS1010, D SHRT E. 49th 11-4-54, PTD aluminum band 6-8-59, OOS 1987 stored Brookpark, sold to Indiana Railroad at Indianapolis, shipped by rail 6-10-88 |
| 217–218 | loaded 10-27-54 on flat cars NYC499477 and CNW-47909, D SHRT E. 49th 11-3-54, PTD aluminum band |

|   |   |
|---|---|
|  | 6-11-59, OOS ACC 5-1982 stored Windermere, transferred to Kingsbury, SCR 1987 |
| ★219–220 | loaded 10-30-54 on flat cars NYC500298 and WM2685, D SHRT E. 49th 11-5-54, first train set to operate over line in 2-1955, PTD aluminum band 6-16-59, ACC side scraped in Windermere yard 5-3-63, poor job of body putty and paint, cracked and peeled, car could be recognized from a distance, fire damage 2-25-69 repaired, OOS 10-1982 with fire damage, stored Windermere, moved to Kingsbury 1987, SCR 1987 |
| 221–222 | loaded 10-31-54 on flat cars T&P5581 and C&O81031, D SHRT E. 49th 11-8-54, PTD aluminum band 6-18-59, OOS 1987 stored at Brookpark, SCR 1988 |
| 223–224 | loaded 10-31-54 on flat cars CNJ167 and RI90808, D SHRT E. 49th 11-9-54, PTD aluminum band 6-24-59, bad anchor plate 1976, OOS 12-1984 stored Brookpark, transferred to Kingsbury 1987, SCR 1987 |
| 225–226 | loaded 11-8-54 on flat cars NYC499544 and NYC-500469, D SHRT E. 49th 11-16-54, PTD aluminum band 6-29-59, PTD 3-29-65, ACC motor dislodged derailing car 226, which was heavily damaged upon smashing a line pole in 11-1974, sold for SCR 1979 |
| 227–228 | loaded 11-15-54 on flat cars NYC499384 and NYC-500354, D SHRT E. 49th 11-22-54, PTD aluminum band 7-2-59, OOS 3-1985 stored Windermere, transferred to Brookpark 1988, SCR 1988 |
| 229–230 | loaded 11-15-54 on flat cars NYC499823 and NYC-499760, D SHRT 11-23-54, PTD aluminum band 7-7-59, 229 damaged in fire at Windermere shop 3-3-59 at 1:00 A.M. when leaking gasoline from motor car 016 caught fire, repaired 4-28-59, OOS 2-1985 stored Windermere, SCR 1987 |
| 231–232 | loaded 11-24-54 on flat cars NYC499780 and C&O-216687, D SHRT E. 49th 12-2-54, PTD aluminum band 7-10-59, PTD 8-26-66, ACC side of car 231 badly scraped on platform in 1976, OOS 1988 stored Brookpark, SCR 1988 |
| 233–234 | loaded 11-24-54 on flat cars THB1833 (Toronto Hamilton and Buffalo) and C&O216543, D SHRT E. 49th 12-1-54, PTD aluminum band 7-15-59, PTD 8-15-66, OOS 1988 stored Brookpark, SCR 1988 |
| 235–236 | loaded 12-8-54 on flat cars NYC499550 and DRGW-22062, D SHRT E. 49th 12-14-54, PTD aluminum band 7-20-59, ACC car 151 hit car 236 at W. 98th Station 11-27-70, car repaired, moved to E. 55th yard 8-1984, OOS 11-1984 stored Kingsbury, SCR 1987 |

| | |
|---|---|
| 237–238 | loaded 12-8-54 on flat cars NYC500374 and IGN8596, D SHRT E. 49th 12-13-54, PTD aluminum band 7-23-59, PTD 8-2-67, OOS 1987 Kingsbury, SCR 1987 |
| ★239–240 | loaded 12-15-54 on flat cars NYC499738 and NYC499589, D SHRT E. 49th, stored Brookpark, transferred to Windermere, transferred to Kingsbury 1987, SCR Windermere 1987 |
| ★241–242 | loaded 12-21-54 on flat cars NYC499833 and PM16722, D SHRT E. 49th 12-27-54, PTD aluminum band 7-31-59, PTD 7-15-66, OOS 1988 stored Brookpark, SCR 5-20-88 |
| 243–244 | loaded 12-24-54 on flat cars ERIE8165 and PM16766, D SHRT E. 49th 12-29-54, PTD aluminum band 8-5-59, bad anchor plate and frame 1976, OOS 1988 stored Brookpark, SCR 1988 |
| 245–246 | loaded 12-31-54 on flat cars NYC500570 and CNW46111, D SHRT E. 49th 1-5-55, PTD aluminum band 8-11-59, PTD 8-1-66, OOS 1987 stored Brookpark, SCR 5-17-88 |
| 247–248 | loaded 1-5-55 on flat cars LV10035 and PRR474270, D SHRT E. 49th 1-11-55, PTD aluminum band 8-18-59, OOS and SCR 8-1984 |
| ★249–250 | loaded 1-13-55 on flat cars THB1859 and CIL7229, D SHRT E. 49th 1-21-55, PTD aluminum band 10-30-59, PTD 8-25-67, bad pinion OOS 11-1984, stored Kingsbury, SCR 1987 |
| ★251–252 | loaded 1-21-55 on flat cars NYC499644 and CNW44427, D SHRT E. 49th 1-27-55, PTD aluminum band 11-5-59, PTD 6-14-66, ACC derailed at West Park Station 1-17-66, cars repaired by 7-1-66, sold for SCR 1981 |
| ★253–254 | loaded 1-21-55 cars PM16578 and CNW44311, D SHRT E. 49th 1-28-55, PTD aluminum band 11-12-59, PTD 6-19-67, bad frame 6-1982, OOS 1-1985 stored Windermere, transferred to Brookpark 1988 and SCR 1988 |
| ★255–256 | loaded 2-2-55 on flat cars NYC500061 and KCS1095, D SHRT E. 49th 2-10-55, PTD aluminum band 11-10-59, ACC car 255 hit Shaker car 87 at CUT 12-1-66, 87 repaired at Windermere 12-1-66, out 12-7-66, 255 PTD 6-26-67, OOS 12-1980, 255 bad frame 6-1982, stored Windermere, SCR 1987 |
| ★257–258 | loaded 2-14-58 and 2-20-58 on flat cars SOU51850 and D&RGW22193, D SHRT E. 49th 2-20-58 and 2-28-58, PTD aluminum band 9-7-62, ACC 1983 car 257 had serious electrical fire, transferred to Kingsbury, SCR 1987 |
| ★259–260 | loaded 2-24-58 and 2-26-58 on flat cars SOU51858 and ATSF93817, D SHRT E. 49th 3-3-58 and 3-6-58, PTD |

|  |  |
|---|---|
|  | aluminum band 11-14-61 and 10-24-62, car 260 destroyed by fire, OOS at Kingsbury 1987, SCR 1987 |
| *261–262 | loaded 3-3-58 and 3-7-58 on flat cars SLSF2156 and C&O81042, D SHRT E. 49th 3-10-58 and 3-14-58, PTD aluminum band 9-19-62, ACC in 1973 anticlimber on 262 badly smashed, sat under repair for several years, then SCR 1984 |
| *263–264 | loaded 3-11-58 and 3-21-58 on flat cars CN663345 and C&O81019, D SHRT E. 49th 3-18-58 and 3-27-58, PTD aluminum band 10-10-62, ACC car 263 hit bumper stop at Brookpark storage yard 9-3-69, OOS 1988 stored Brookpark, SCR 1988 |
| *265–266 | loaded 3-20-58 and 3-28-58 on flat cars CRR10114 and MILW66321, D SHRT E. 49th 3-26-58 and 3-28-58, PTD aluminum band 9-13-62, OOS 1987, sold to NORM |
| *267–268 | loaded 4-1-58 and 16-58 on flat cars CN663239 and SP142953, D SHRT E. 49th 4-7-58 and 4-22-58, PTD aluminum band 10-3-62, ACC at airport station in 1977 end of car 268 crushed against bumper post in station, SCR 1984 |
| *269–270 | loaded 4-23-58 and 5-13-58 on flat cars SP142892 and IC60031, D SHRT E. 49th 4-29-58 and 5-19-58, PTD aluminum band 9-26-62, 269 had bad frame 1977, OOS, sold for SCR 1981 |
| 204–257 | remated 9-1984 and never used |
| 248–259 | remated 9-1984 and never used |
| 247–260 | remated and entered service with old numbers, REN 8-31-84 to 271–272, OOS 1987 stored Brookpark, SCR 1988 |
| 203–258 | remated and entered service with old numbers, REN 8-31-84 to 273–274, OOS 3-1985 stored Windermere, transferred to Brookpark, SCR 1988, motorman's cab of 274 (258) owned by Blaine Hays |

### Fleet Statistics: State-of-the-Art (SOAC) Cars

BLDR: St. Louis Car Co. 1972; LEN: 75'; seats: #1 low-density car 64, #2 high-density car 72; MOT: truck mounted 175 hp DC; TRKS: General Steel Industries; CTL: Garrett AiResearch solid state chopper with thyristors; W: 90,000 lbs. Two cars operated as a train in demonstration trips over American high-platform rapid transit systems. The mock-up was in Cleveland at Public Square and at Parmatown and Severance shopping centers in 1-1975, and cars 1 and 2 operated over the CTS rapid transit for two weeks in 2-1975 and 3-1975. Cleveland was the only city where pantographs were used. All other demonstrations used third-rail shoes. After the demonstration and a stint on the New York City subway system, the cars were acquired by Seashore Trolley Museum 1990.

SOAC (mock)
SOAC 1           low-density car
SOAC 2           high-density car

### Fleet Statistics: Pullman "Airporter" Stainless Steel Cars, 151–180

BLDR: Pullman-Standard, 151–170 (F 04) D 10 and 11 1967, 171–180 (F 06) D 11 and 12 1970; LEN: 70'; W: 64,000 lbs; seats: 80; MOT: 4GE100 hp number 1250; CTL: GE Motor-operated camshaft with static regulation; cars had red- and orange-colored fluorescent tape used for trim, fiberglass ends. Cars 171–180 had smaller front windows; cars 161–170 were shipped on their own wheels in a Norfolk & Western train from Chicago Pullman plant to E. 32d SHRT siding on 10-31-67. Idler cars N&W 13107 and N&W 13244 had dual couplers for this purpose. Records are missing, but other 20 Pullman "Airporters" were delivered the same way. Cost of F 06 $251,950 each. All cars OOS and stored Brookpark yard, except 172, which is operable and at E. 55th yard. Known by employees as "B" cars throughout the years.

151   ACC 11-27-70, hit rear of car 236 at W. 98th Station, extensive work required, car in Windermere shop from 11-27-70 to 12-9-70, REB for Westinghouse pulse width modulation inverter testing 1972, received DC motors and green/white/black paint and carpeted interior, carried banners proclaiming "Rapid RTA 35¢" in 1975, fire OOS 1982, stored inside Brookpark shop

152   renovated for Westinghouse pulse width modulation inverter testing 1972, received AC motors and green/white/black paint, carried banners proclaiming "Rapid RTA 35¢" in 1975, OOS ACC 1-1985

153   renovated for Westinghouse pulse width modulation inverter testing 1972, received AC MOT and green/white/black paint and carpeted interior, carried banners proclaiming "Rapid RTA 35¢" in 1975, fire OOS 5-1982

154   pilot car for pulse width modulation demonstration, instrumentation-filled car, all seats removed, received AC MOT, ACC fiberglass end smashed 1974, repaired, carried banners proclaiming "Rapid RTA 35¢" in 1975

CTS Airporter 157, Pullman-Standard, 1967. *(Pullman-Standard photo, Blaine Hays Collection)*

| | |
|---|---|
| 155 | ACC side bruised 1975, repaired |
| 160 | used in mock "fire evacuation from tunnels" exercise with Brookpark Fire Department in 1979 |
| 161 | reactivated for special run 11-15-88 for twentieth anniversary of airport service |
| 162 | fire 10-1982, OOS and stored |
| 163 | ACC 1976, side of car bruised and frame heavily damaged, repaired at GE E. 49th Street plant, CTL problems, OOS 1982, repaired and used as tow car at Brookpark yards in 1980s |
| 164 | ACC front end of 164 hit rear end of 169 at W. 34th Street on 9-12-68, ACC front pushed in 1976 and OOS |
| 165 | ACC bad coupler boxes, OOS 1-1982 |
| 167 | front-end ACC, stored at Kingsbury spring 1985, moved to Brookpark 1992 |
| 169 | Vandalism to roof 12-4-67 and repaired, ACC W. 34th Street, car 164 hit rear of 169, had end PTD 12-17-68, experiment |

The Pullman Airporters offered spacious interiors and featured luggage racks at both front and rear doors. *(Cleveland Transit photo, Blaine Hays Collection)*

       done 6-12-69 by Westinghouse and Reliance Electric to install test propulsion equipment, stored at Windermere spring 1983, moved to Kingsbury, last car to operate over Kingsbury yard tracks, had tree growing up through body which had to be cut for car to be moved, sent to Brookpark 1992

170    destroyed in electrical fire at airport station 11-11-72, fire so intense that stainless steel was bowed, SCR 1-1973, parts remained in Windermere yard for years

171    struck by lightning in 1972 and spent so much time in the shop it acquired a rollsign that read "Windermere Shop Express"

172    reactivated for special run 11-15-88 for twentieth anniversary of airport service

174    ACC damage 1976 and OOS

178    ACC destroyed fiberglass end and part of frame in 1976

179    ACC damage 1976

180    derailed at high speed, smashing into center of cantenary support at West Park Station 1978, stored Brookpark for years, moved to Windermere to be picked to the bone so other cars could live, SCR fall 1983

### Fleet Statistics: Tokyu High-Platform Cars, 181–200 and 301–340

BLDR: Tokyu Car Corp., Yokohama, Japan; LEN: 75'; seats: 84; W: 76,000 lbs; MOT: GE forced ventilation; CTL: GE motor-operated camshaft; cost: $933,000 each; cars delivered with "Mentor" rollsigns.

All cars accepted between 8-30-84 and 9-30-85. Some cars in this fleet were involved in a sale-leaseback agreement whereby equity interest was purchased by private firms for income tax purposes, resulting in RTA receiving money from firms and the firms getting a tax break.

180 (mock)    no trucks or controller, consists of a third of fully built car body, stored at RTA Central Rail shops and picked for parts

188    sent to IX Center over Conrail tracks 3-15-90, and displayed there to show feasibility of extending rapid transit service to the center

192, 198    known as backhoe cars; car 198 hit a construction backhoe, damaging front of car, backhoe then pivoted striking center of both cars, and then struck rear of 192, shop forces put backhoe hash marks on cars

301    first Tokyu car involved in ACC 5-1985 at W. 98th station, car OOS

Balloon    air-balloon Tokyo car, used by RTA for promotions

## SECTION 3
### Rapid Transit Service Cars

In addition to the cars listed below, several passenger cars were also CON to service cars, and these are noted in the roster above: 3, 10, 011(2), 012(2), 15(2), 031, 1233–1235, 2318–2319, 2365.

OX    Box Freight Motor, BLT G. C. Kuhlman 1924 for Eastern Michigan Electric as number 200 (this might have been 202), sold to Northern Ohio Traction & Light 1929 and REN 1078, sold to CIRR along with flat car X12 1932, brought to Cleveland under own power over Northern Ohio interurban. REN X78, REN by SHRT to 78, then REN OX in 1948 upon arrival of Pullman PCC cars of which there was a "78." The "OX" name implied that the car did the work of an ox on the line. CON to line car by addition of movable roof platform. A claim to fame for OX was its use in bringing Santa Claus to waiting children on Shaker Square for the day-after-Thanksgiving tree lighting ceremony remembered by thousands of children over several decades. One night in 1976 roof platform was accidentally left sticking out, caught the edge of Triskett Station, pulling it off its frame. Sold to Trolleyville, USA, in 1984 and moved there in 1985

X12    Flat Car (high platform), BLT Northern Ohio Traction & Light 1915 as DT flat car with MCB air couplers, sold to NORM 1984, SCR by museum for parts

CIRR 78, FN 1078 Northern Ohio Traction freight motor, REN OX by SHRT, G. C. Kuhlman, 1924. *(Blaine Hays Collection)*

101    Line Car, BLT by Detroit, Plymouth & Northville predecessor Detroit, Jackson & Chicago Railway as #77, components used by Detroit United Railways to construct line car 7763 in 1923 for Eastern Michigan Railways, Peckham TRKS, sold 1932 to CIRR, shipped to Cleveland over Lake Shore Electric, REN 101, sold by SHRT to Ohio Railway Museum 1965

016(2)    Buda gasoline work motor, purchased by CTS from CUT 1955 for $300, FN X54, caught fire 3-3-59 at Windermere shop when gasoline leaked from the car causing severe damage, sold to SHRT for $300 9-1964, fixed to operating condition, transferred to RTA, still in use

017(2)    Enclosed Flat Car, purchased by CTS from CUT 1955, for $350, FN X61, car severely damaged with 016 on 3-3-59 at Windermere shop, top SCR 1959, remainder CON to flat car

018(2)    Tower Car, BLT J. R. McCardell and Co., NJ, FN X71, purchased by CTS from CUT 1955, sold to SHRT for $350 9-1964, ACQ by Trolleyville, USA, 1982, now at NORM

019(2) Tower Car, BLT J. R. McCardle and Co., NJ, FN X72, purchased by CTS from CUT 1955, sold to SHRT 9-1964 for $325, ACQ by NORM 1982

020(2) Wire Reel Car, FN X63, purchased by CTS from CUT 1955 for $90, destroyed by the side of the tracks at Brookpark yards in 1984

020(3) Tamper, BLT Plasser-American 1983

021(2) Flat Car with Tower, BLT by CTS for rapid transit use, authorized for SCR 9-21-61, presumed SCR

021(3) Tamper, BLT by Plasser-American 1974, serial #2391

022 Flat Car, four wheel, FN X62, purchased by CTS from CUT 1955 for $40, still used

023 Flat Car, four wheel, FN X64, purchased by CTS from CUT 1955 for $40, authorized for scrap 9-21-61, presumed SCR

★ unnumbered flat car, four-wheel BLT by Fairmount, not included in inventory

024(5) Line Car, FN (1217) 17 and components from (1226) 26, car arrived at Windermere 4-1960, construction carried out by CTS shop forces during 1960, arch bar TRKS and pantograph installed. Long-time, well-liked motorman of car was Clarence "Whity" Hofstrom; to this day PTD on MU control door inside car is the following from Whity's retirement: "Goody Day—Whity last day 6-22-72." (A thrill of a lifetime was for new CTS employee Blaine Hays to get in cab of 024 and have Whity place his hand over his on the controller and allow him to operate the car from Quonset building on the west side to E. 49th St. siding.) PTD to RTA colors 1978

032 Line Car, FN 201, BLT Fuji Heavy Industries for RTA 1980, $764,000, for use on SHRT, diesel-powered car spent much time idle because of engine problems, towed much of the time by Locomotive 033, engine repaired 1996

033 65-ton diesel locomotive, BLT GE 7-1943 as U.S. Naval Ammunition Depot number 65-00056 at Puget Sound, WA, later sold to Alaska Lumber and Pulp Company at Sitka, sometime after 1977 returned to GE in Cleveland to be REB, ACQ by RTA 1980

0137(2) Snow Sweeper, DT, transferred to CTS Rapid from streetcar system 1953, PTD dark-green and, later, yellow, equipped with heavy steel plow on rapid (which broke car's wooden frame), damaged by fire on 3-3-59 at Windermere shop and determined to be no longer needed, retired 8-22-60 (also reported retired 9-22-60), parts sold to Connecticut Electric Railway Museum on 10-4-60.

0308 Motor Rail Car, BLT Geismar 1988

| | |
|---|---|
| 0516 | Dump Freight, DT, transferred to CTS Rapid from streetcar system 1953, frame bent unloading ballast, MOT removed and body sold to NORM 1979, SCR by museum at Kingsbury for parts |
| 0518 | Dump Freight, DT, transferred to CTS Rapid from streetcar system 1953, MOT removed and car sold by RTA to NORM 1979, moved 12-1979 to museum |
| 0610 | Flat Car, DT, sold to SHRT in lieu of 4000-series passenger cars, shipped to SHRT 11-4-53, sold to NORM 1981 |
| 0611 | Flat Car, DT, transferred to CTS Rapid from streetcar system 1953, sold to NORM 1981 and moved to museum |
| 0615 | Flat Car, DT, transferred to CTS Rapid from streetcar system 1953, sold to NORM 1981 and moved to museum |
| 0616 | Flat Car, 50 ton, LEN 51', W 45,150 lbs, BLT Maxson Corp. 2-1979 |
| 0617 | Flat Car, 50 ton, LEN 51', W 45,150 lbs, BLT Maxson Corp. 3-1979 |
| 0618 | Dump Freight Trailer, 50 ton, LEN 51', W 80,300 lbs, BLT Maxson Corp. 8-1979 |
| 0619 | Hopper Car, 50 ton, LEN 51' W 50,000 lbs, BLT Maxson Corp. 5-29-1979 |

RTA 0619, hopper car, Maxson Corporation, 1979. *(Maxson Corp. photo, Blaine Hays Collection)*

| | |
|---|---|
| 0710 | Crane Car, DT, renovated at Harvard 11-1953, sold to SHRT in lieu of 4000-series passenger cars, shipped to SHRT 2-1-54, main gear sheared in 1978, CON to motorized flat had "no springs-honest weight" on front, sold to NORM 1981 |

0711  Crane Car, DT, renovated at Harvard and transferred to CTS Rapid from streetcar system 3-10-53, ACQ TRK from 0710 to continue in service, ACQ by NORM 11-1980

0800  Rail Scrubber, ST, transferred to CTS Rapid construction project at SHRT Kingsbury Yard from Denison Station 1953, retired 5-17-56, SCR with components sold to Illinois Railway Museum 8-2-57 after CTS rapid construction completed

R637  Box Freight Motor, Cincinnati & Lake Erie, sold to American Aggregates 1939, ACQ by Ron Jedlicka and moved to Kingsbury shops for storage, moved off property 1980

GE77  Caboose (wooden), REB 1982 GE, sold to RTA for work train, never used and set afire by arsonists 1988, frame SCR 1989

109  PCC Rapid Transit Car, CON to garbage hauling car 2-1990 by removal of seats, PTD white/orange/red RTA scheme, PTD 10-1990 into as-delivered CTS blue/silver/red livery with replica of original CTS winged logo and reinstallation of seats for display at opening of new joint Rapid Transit concourse at Tower City 12-17-90

ROSTER SHORT

*Car Numbering Has Historic Roots:* In 1979 RTA equipment administrator Chuck Kocour asked Blaine Hays to organize the numbering of RTA work cars and miscellaneous equipment. Hays did so with historical continuity in mind. That is why the newer RTA-era work cars are numbered consecutively, as a carry-over from the CR numbering system started prior to the turn of the century. For example, flat car 0616 built in 1979 and currently in use on the RTA rapid transit system is preceeded by flat car 0615 built in 1928, now an exhibit at NORM. Transit history lives on!

PART C

## *Roster of Trackless Trolley Passenger Equipment*

The question "When did the last trolley run?" often comes up. The response of January 24, 1954, is sometimes met with, "No, I remember trolleys into the 1960s." The confusion results from the term "trolley" meaning either the streetcars or the twin-poled electric buses, which were always known in Cleveland as trackless trolleys and which operated on some east side lines until June 15, 1963. Completely apart from the streetcars, Cleveland had an extensive electric bus network with operations beginning March 1, 1936.

TT 1171, St. Louis Car, 1949, at West 9th Street and Superior Avenue. *(Anthony F. Krisak photo, Richard Krisak Collection)*

Adoption of this mode began because of a track replacement ultimatum from the City of Cleveland, not from operational considerations. The city announced its intention to repave Hough Avenue, and after a study of the transit routes between Euclid and Superior avenues, Cleveland Railway decided to convert to trackless the Hough Avenue line and to reroute the Wade Park streetcar line.

Successful experience with trackless trolley operation over this line resulted in ordering additional units, but not in expanding the mode. The reason for this reluctance was that as the time for renewal of the company's operating franchise (the Tayler Grant) came closer, changes to the system became more uncertain. Expansion of the trackless system did not resume until after World War II. It then resulted in fifteen electric bus routes operating under 600-volt DC wires and using 461 vehicles.

The St. Louis, GE, and Westinghouse fleets were mechanically different (1000–1074, 1075–1174). They were usually kept at different stations and not mixed on lines.

### Fleet Statistics: Pullman TT 900–919, Fleet 36

BLDR: Pullman-Standard 1936; ACQ: 1936; model: 35J-100-40C.X; LEN: 33' 6"; seats: 40; W: 20,300 lbs; CTL: GE P.C.19; MOT: GE 1154; front and center doors. These were the first trackless trolleys in Cleveland, D between 2-8-36 (coach 900) and 3-16-36. Trolleys 900–927 were assigned to the Hough line, and in the late 1940s became rush-hour extras on east side lines. All OOS 9-16-54; 901 SCR 11-29-55; 916 sold to E. M. Sargi 3-10-55; all others sold for SCR 1955.

### Fleet Statistics: Pullman TT 920–927, Fleet 36

BLDR: Pullman-Standard 1937; ACQ: 1937; model: 37FD-100-40 C.X.; LEN: 33' 6"; seats: 40; W: 20,800 lbs; CTL: GE P.C.19; MOT: GE 1154; front and center doors. Hough line ridership soared upon

TT 920, Pullman-Standard, 1937. *(Cleveland Railway photo, Blaine Hays Collection)*

conversion to TTs, and these eight additional units were ordered to meet the needs. D between 6-19-37 and 7-6-37; all OOS on 9-16-54, and 925 was sold to Mr. Fleck at Twinsburg Sportsman Club 12-1954; all others sold for SCR 1955.

### Fleet Statistics: Super Twin TT 6000, the Queen Mary

BLDR: Twin Coach 1940; ACQ: 1940; model: 58G TT; LEN: 47' 6"; seats: 56; MOT: WH 1-1442-A; W: 26,000 lbs; front and center doors (behind articulation)

6000    D 6-26-40 as a six-wheel demonstration articulated TT, dubbed the Super Twin, Cleveland's Queen Mary. The rear section was designated as a smoking section in an era when such a thing was fashionable. ACQ 1942 and operated exclusively on Hough Line. It sat in Hough Station for long periods awaiting parts due to its one-of-a-kind status. OOS 2-3-54, it found its way to Harvard Shops. On 6-8-54 its status was changed to service equipment and used to test weather conditions and colors for paint on the new rapid transit cars. During this period it appeared frequently in variations of strange color patterns. Sold to City Auto Wrecking 3-20-57. The Queen Mary was often used to pick up students from Thomas Jefferson Junior High School—as long as the qualified operator was on duty.

TT 990, Pullman-Standard, 1945. *(Cleveland Transit photo, Blaine Hays Collection)*

*Fleet Statistics: Pullman TT 950–999, Fleet 35*

BLDR: Pullman-Standard 1945–46; ACQ: 1945–46; model: 44; LEN: 37' 11"; seats: 44; W: 23,200 lbs; CTL: GE MRC automatic; MOT: GE 1213 D7 compound; front and center doors. The first CTS-ordered TTs, ACQ for system expansion to Cedar and Wade Park. D between 12-1945 and 6-1946; all OOS 1957–58 and sold for SCR; 982–992 OOS at Superior Station in 1955 and SCR 1957–58. A portion of this fleet purchased under first series equipment trust 6-1-46.

*Fleet Statistics: St. Louis TT 1000–1074, Fleet 33*

BLDR: St. Louis Car 1947–48; ACQ: 1947–48; model: STL44; LEN: 37' 8"; seats: 44; MOT: GE; W: 23,320 lbs; front and center doors. A portion of this fleet purchased under first-series equipment trust dated 6-1-46; another purchased under series two dated 12-15-46. As TT lines were abandoned, vehicles were stored as follows:

TT 1000, St. Louis Car, 1947. *(Cleveland Transit photo, Blaine Hays Collection)*

Lorain Station: 1009, 1016, 1018, 1024, 1026, 1029, 1031, 1033, 1034, 1038, 1042, 1043, 1046, 1048, 1052, 1058, 1060
Superior Station: 1079, 1084 (dead storage), 1120, 1133.

In 1961 the following vehicles were moved: 1031, 1039, 1040, 1051, 1059, 1062, 1064, 1069, 1072 to Woodhill Station; 1023, 1083, 1091, 1117, 1150, 1161, 1169 to Superior Station. They were sold for scrap as follows:

1000, 1019, 1021, 1044–1049, 1057–1059, 1065–1070, 1073–1074 • to City Auto Wrecking 1958
1041–1043, 1050, 1055 • to Kubentz Iron 1958
1052 to J. Ours and moved to New Brighton, PA, 1963; D 12-1947, for sale 4-18-63, sold 6-26-63. Selected as a storage shed, it thereby escaped scrapping. Seats were removed and bins installed, but by 1983 it was transferred to A.&B. Salvage Co. in Lancaster Twp., where it was placed in view of Interstate 79. Greg Walz of Pennsylvania Trolley Museum spotted it and informed Bob Humanchuk of NORM, who made an inspection, found it sound, obtained it, and then donated it to Seashore Trolley Museum. On 7-7-90 1052 left Pennsylvania for Maine and took its place in line for restoration. It is the only known surviving example of a St. Louis trackless trolley of its type.
1053 painted for "Safe Crossing" in 1952
1060 serious front-end ACC
1071 OOS (ACC) 1-29-58, sold 8-28-58

### Fleet Statistics: Marmon TT, 1200–1274, Fleet 34

BLDR: Marmon-Herrington Co. 1947–48; ACQ: 1947–48; model: TC44; LEN: 36' 5"; seats: 44; MOT: GE; W: 17,320 lbs; front and center doors; first D (1200) 11-10-47; last D (1274) 3-2-48. TTs 1225–1274 were sold by CTS directly to Mexico City in 8-1959 (REN 3725–3774) and were loaded during 9-1959 and 10-1959. Others were sold as follows:

1203, 1206, 1210, 1215 • to Rearich 1966
1214, 1219–1220, 1223 • to Transport Parts 1966
1200 Transport Parts 10-1966, resold to Mexico REN 3717
1201 Toronto 4-26-63, loaded 5-24-63 REN 9145, SCR 1972
1202 Transport Parts 10-1966, resold to Mexico REN 3718
1204 Transport Parts 10-1966, resold to Mexico REN 3719
1205 Toronto 4-26-63, loaded 5-24-63 REN 9146, SCR 1972

Marmon 1308 typifies Cleveland's trackless trolley fleet. Best remembered is the rivet-covered interior. *(Cleveland Transit photo, Blaine Hays Collection)*

| | |
|---|---|
| 1207 | Toronto 4-26-63, loaded 5-24-63 REN 9147, SCR 1972 |
| 1208 | Transport Parts 10-1966, resold to Mexico REN 3720 |
| 1209 | Toronto 4-26-63, loaded 5-24-63 REN 9148, SCR 1972 |
| 1211 | Toronto 4-26-63, loaded 5-24-63 REN 9149, SCR 1972 |
| 1212 | Toronto 4-26-63, loaded 5-24-63 REN 9150, SCR 1972 |
| 1213 | fire damage, sold Transport Parts 10-1966 |
| 1216 | Transport Parts 10-1966, resold to Mexico REN 3721 |
| 1217 | Transport Parts 10-1966, resold to Mexico REN 3722 |
| 1218 | Toronto 4-26-63, loaded 5-24-63 REN 9151, SCR 1972 |
| 1221 | Transport Parts 10-1966, resold to Mexico REN 3723 |
| 1222 | Transport Parts 10-1966, resold to Mexico REN 3724 |
| 1224 | Toronto 4-26-63, loaded 5-24-63 REN 9152, SCR 1972 |

*Note:* This fleet purchased under second-series equipment trust 12-15-46.

### Fleet Statistics: St. Louis TT 1075–1174, Fleet 39

BLDR: St. Louis Car Co. 1948–49; in two separate deliveries, ACQ: 1948–49; model: STL44; LEN: 37' 8"; seats: 44; MOT: WH; W: 22,840 lbs; front and center doors; first D (1075) 1-19-48; final D (1129) 3-11-48: first D of second shipment (1130) 4-11-49; final D (1174) 7-5-49. 1127 and 1137 were OOS at Superior Station in 1954 awaiting repair from fire damage. Sold for scrap as follows:

1088, 1101–1103, 1116, 1133, 1140, 1144, 1174 • City Auto Wrecking 1958

1095, 1121, 1130–1132, 1135, 1138–1139 • Kubentz Iron 1958

TT 1157, St. Louis Car, 1949. *(Cleveland Transit photo, Blaine Hays Collection)*

| | |
|---|---|
| 1113 | J. Ours 1960 |
| 1136–1137, 1146 | • Kricfalusy 1963 |
| 1092 | ACC (fire) OOS 3-18-58, sold 8-28-58 |
| 1110 | ACC 5-11-58, sold 8-28-58 |
| 1123 | ACC OOS 8-11-58, sold City Auto Wrecking |
| 1161 | had serious electrical fire |
| 1169 | extensive fire damage, OOS 4-26-62, sold 8-15-62 |

*Note:* This fleet purchased under second-series equipment trust 12-15-46.

### Fleet Statistics: Brill TT 800–831, Fleet 38

BLDR: J. G. Brill 1937; ACQ: 1951; LEN: 37' 10"; seats: 40; MOT: WH T40; front and center doors. ACQ from Louisville Railway Co. in 1951 and REN for CTS service. During their short life in Cleveland, they were used only as trippers and mostly on the westside lines. All OOS 2-15-55 except 815, which suffered mechanical problems and

TT 801, former 224 of Louisville Railway, J. G. Brill, 1936, acquired by CTS, 1951. *(Cleveland Transit photo, Blaine Hays Collection)*

was OOS 2-2-54 and SCR 6-25-54. All sold for SCR 1955. They were OOS and REN as follows:

800–807, 810–814, 816–826 • OOS Superior Station 1954
808–809, 827–831 • OOS St. Clair Station 1954

| | |
|---|---|
| 800 | FN 221 |
| 801 | FN 224 |
| 802 | FN 219 |
| 803 | FN 217 |
| 804 | FN 229 |
| 805 | FN 232 |
| 806 | FN 218 |
| 807 | FN 206 |
| 808 | FN 213 |
| 809 | FN 208 |
| 810 | FN 225 |
| 811 | FN 230 |
| 812 | FN 223 |
| 813 | FN 202 |
| 814 | FN 207 |
| 815 | FN 216 |
| 816 | FN 211 |
| 817 | FN 228 |
| 818 | FN 227 |
| 819 | FN 212 |
| 820 | FN 214 |
| 821 | FN 210 |

| | |
|---|---|
| 822 | FN 220 |
| 823 | FN 203 |
| 824 | FN 231 |
| 825 | FN 226 |
| 826 | FN 205 |
| 827 | FN 209 |
| 828 | FN 215 |
| 829 | FN 201 |
| 830 | FN 222 |
| 831 | FN 224 |

### Fleet Statistics: Marmon TT 1275–1324, Fleet 37

BLDR: Marmon-Herrington Co. 1951; ACQ 1951; model: TC49; LEN: 40' 8"; seats: 50; MOT: WH; W: 20,060 lbs; front and center

TT 1275, Marmon-Herrington, 1951. *(Cleveland Transit photo, Blaine Hays Collection)*

doors. This fleet was known as the "Big Marmons" as opposed to 1200–1274, which seated six fewer persons and was Cleveland's last fleet of TTs. First D (1275) 9-4-51; last D (1323) 10-30-51. TT 1324 was shipped to Brussels to serve as an exhibit at the Brussels Automobile Show in January 1952; it was returned and entered CTS service 6-5-52. All OOS 6-63 and sold for scrap as follows:

1275, 1277, 1280–1281, 1285, 1288, 1290, 1292, 1296–1297, 1300 • Rearich 1966
1276, 1278–1279, 1282–1284, 1286, 1287, 1289, 1291, 1293–1294, 1298, 1301–1303, 1305, 1310–1312, 1314–1323, 1324 • Transport Parts 1966
1295, 1306, 1309, 1313 • Kricfalusy 1966
1299, 1304, 1307–1308 • Kubentz Iron 1966

*Note:* This fleet was purchased for service on St. Clair but was used on E. 105th St., replacing smaller TTs on that line. These TTs were also used on weekends on other lines that ran from Woodhill Station.

### *Fleet Statistics: Pullman TT 850–882, Fleet 40*

BLDR: Pullman-Standard 1947–48; ACQ: 1952–53; model: 45; LEN: 37' 10"; seats: 45; MOT: WH T40; W: 23,980 lbs; front and center doors. ACQ from United Transit System of Providence, RI with D starting 12-29-52 and completed 2-13-53. In the roster the first number is Cleveland, followed by United Electric number, then United Transit when United Electric became United Transit. The fourth number is the year built.

850, 1351, 7501, 1947, D 2-11-53, OOS Superior 2-24-59, sold 5-25-59
851, 1352, 7502, 1947, D 1-30-53, OOS 11-21-58, sold Herold Respanand
852, 1353, 7503, 1947, D 1-30-53, OOS 10-27-58, sold Kricfalusy
853, 1354, 7504, 1947, D 1-29-53, OOS Superior 4-26-62, sold 8-15-62
854, 1355, 7505, 1947, D 1-30-53, OOS 9-3-58, sold Kricfalusy
855, 1356, 7506, 1947, D 1-30-53, OOS 11-21-58, sold Kricfalusy
856, 1382, 8501, 1948, D 1-21-53, OOS 1-21-58, sold Kricfalusy
857, 1383, 8502, 1948, D 1-21-53, OOS 10-27-58, sold Kricfalusy
858, 1385, 8504, 1948, D 1-26-53, OOS (ACC) 5-2-55, sold 6-19-56
859, 1387, 8506, 1948, D 1-26-53, OOS 10-27-58, sold Kricfalusy
860, 1389, 8508, 1948, D 1-26-53, OOS 9-3-58, sold Kricfalusy
861, 1390, 8509, 1948, D 1-26-53, OOS 3-19-59, sold 5-25-59
862, 1393, 8512, 1948, D 1-26-53, OOS 10-27-58, sold Kricfalusy
863, 1407, 8526, 1948, D 1-27-53, OOS 10-27-58, sold City Auto Wrecking

864, 1408, 8527, 1948, D 1-28-53, OOS 9-3-58, sold Kricfalusy
865, 1409, 8528, 1948, D 1-14-53, OOS 10-27-58, sold City Auto Wrecking
866, 1410, 8529, 1948, D 1-21-53, OOS 9-3-58, sold Kubentz
867, 1411, 8530, 1948, D 2-2-53, OOS 10-27-58, sold City Auto Wrecking
868, 1412, 8531, 1948, D 2-2-53, OOS Superior 1954, OOS 9-3-58, sold Kubentz
869, 1413, 8532, 1948, D 2-2-53, OOS 10-27-58, sold City Auto Wrecking
870, 1414, 8533, 1948, D 2-3-53, OOS 10-27-58, sold City Auto Wrecking
871, 1415, 8534, 1948, D 2-6-53, OOS 10-27-58, sold City Auto Wrecking
872, 1416, 8535, 1948, D 2-3-53, OOS Superior 1954, OOS (ACC) 3-26-56, sold 6-19-56
873, 1417, 8536, 1948, D 2-6-53, OOS Superior 1954, OOS 9-3-58, sold Kubentz
874, 1418, 8537, 1948, D 2-6-53, OOS 9-3-58, sold Omnibus, Chicago 6-16-63, operated on a charter for fans on last day of trackless trolleys in Cleveland 6-15-63, preserved in operating condition at Illinois Railway Museum
875, 1419, 8538, 1948, D 1-19-53, OOS 4-18-63, sold 6-26-63
876, 1420, 8539, 1948, D 2-6-53, OOS (ACC) 8-11-58, sold Kubentz
877, 1421, 8540, 1948, D 2-11-53, OOS 4-18-63, sold 6-26-63
878, 1422, 8541, 1948, D 2-10-53, OOS 3-19-59, sold 5-25-59
879, 1423, 8542, 1948, D 2-13-53, OOS 3-19-59, sold
880, 1424, 8543, 1948, D 2-13-53, OOS 3-19-59, sold
881, 1425, 8544, 1948, D 1-21-53, OOS 3-19-59, sold Kubentz 6-1959
882, 1426, 8545, 1948, D 1-21-53, OOS 4-18-63, sold

*Note:* This fleet was purchased to convert the Superior line from 4000-type streetcars to TTs, the PCCs cars already having been sold to Toronto.

### Fleet Statistics: Pullman TT 883–899, Fleet 41

BLDR: Pullman-Standard 1947–48; ACQ: 1952–53; model: 45; LEN: 37' 10"; seats: 45; MOT: GE; W: 22,470 lbs; front and center doors. Other information also applies to these coaches for 850–882 above.

883, 1345, 7401, 1947, D 1-6-53, OOS 7-7-59, sold J. Ours 8-17-59
884, 1346, 7402, 1947, D 1-2-53, OOS 4-18-63, sold
885, 1347, 7403, 1947, D 1-13-53, OOS (ACC) 9-20-57, sold 5-15-58

TT 888, former 1350 of United Transit, Providence, Rhode Island, Pullman-Standard, 1947, acquired by CTS, 1953. *(Anthony F. Krisak photo, Blaine Hays Collection)*

886, 1348, 7404, 1947, D 1-9-53, OOS 10-27-58, sold City Auto Wrecking

887, 1349, 7405, 1947, D 1-9-53, OOS 11-20-58, sold Kubentz

888, 1350, 7406, 1947, D 1-7-53, OOS Denison 1957, for sale 10-27-58, sold Kricfalusy

889, 1366, 8404, 1948, D 1-14-53, OOS Superior 1957, for sale 4-18-63, sold

890, 1370, 8408, 1948, D 1-13-53, OOS 10-27-58, sold City Auto Wrecking

891, 1377, 8415, 1948, D 1-13-53, OOS Superior 4-26-62, sold 8-15-62

892, 1430, 9404, 1949, D 1-12-53, OOS 4-18-63, sold

893, 1436, 9410, 1949, D 12-29-52, OOS 4-18-63, sold

894, 1490, 9414, 1949, D 1-16-53, OOS 4-18-63, sold

895, 1441, 9415, 1949, D 12-29-52, OOS 4-18-63, sold

896, 1448, 9422, 1949, D 12-29-52, OOS (ACC) 12-9-57 sold 2-3-58

897, 1449, 9423, 1949, D 12-29-52, OOS Superior 4-26-62, sold 8-15-62

898, 1450, 9424, 1949, D 1-8-53, OOS 4-18-63, sold

899, 1456, 9430, 1949, D 1-6-53, OOS Superior 4-26-62, sold 8-15-62

ROSTER SHORTS: "WAYWARD TRACKLESS TROLLEYS"

1. A story is told of a bus driver who worked out of the old Reed Garage on E. 93d Street who drove a bus some days and relieved a run on a trackless trolley on others. In one incident he forgot he

Derelict trackless trolleys at Superior Station testify to the postwar changeover in CTS operations from the age of electricity to the age of the internal combustion engine. *(Cleveland Transit photo, Blaine Hays Collection)*

was driving an electric rather than a gasoline-powered bus, and in his haste to return to the garage he drove right past the end of the wires, pulled off both poles, and coasted several blocks before realizing his embarassing predicament.

2. During the job of cutting down the overhead wires after the discontinuance of TTs, another embarassing situation developed. At the loop of the 14-Kinsman line was a large building the trolleys would encircle. The crew was diligently cutting wire around the building only to discover an errant TT car left stranded behind!

PART D

## Roster of Cleveland Motor Coach Equipment (Passenger)

In April 1925 Cleveland City Council passed a resolution authorizing Cleveland Railway to operate bus routes in the city and out to adjacent suburbs. The company's Morse Rew was named superintendent of transportation and L. H. Vaughn assistant superintendent. The first formal bus operation was the downtown loop, which began on August 17, 1925, but there were two experimental operations of buses in the city prior to that.

In 1904 Cleveland Electric Railway Company tried a twenty-six-passenger chain-driven bus built by Murray as a downtown circulator and for charter use. Nicknamed the "Opera Bus," it operated for only a short time and was not considered a "transit bus."

In 1914 three hood-in-front coaches numbered 50–52 were purchased for feeder operations in areas with difficult terrain or where investment in rail was not justified. One such application was Noble Road hill from Euclid Avenue to the entrance of General Electric's Nela Park. The dominating philosophy of Cleveland Railway was operating rail services, and this prevented any further development of bus routes until March 1, 1924, when the People's Motor Bus Company made application to the state Public Utilities Commission (PUC) for authorization to operate buses. This action was followed in 1925 by a second application from Metropolitan Motor Bus Company, and Cleveland Railway felt pressured to take action. Using its considerable influence in the community, the company forced regulation of city motor transportation away from the PUC through the state legislature's enactment of the Collister-Krueger bill.

To hush criticism and prevent erosion of its public image, Cleveland Railway immediately organized its Motor Coach Division, which featured a distinguishing diamond-shaped insignia. This new division established eight bus routes within the city by the end of 1925.

This is a historical document about the service and fate of the vehicles concerned. Unlike the streetcars, trackless trolleys, and rapid transit cars, buses always had a healthy resale market. Historians use this information to trace the existence of vehicles. Therefore, in addition to the technical fleet information, where known the disposition and date of sale are included, as well as vehicle differences. Buses belonging to other Cleveland operators are mentioned only as they pertain to Cleveland Railway, Cleveland Transit System, or Regional Transit Authority ownership even if they were never operated directly by them.

### *Fleet Statistics: Murray "Opera Bus"*

BLDR: believed to be Murray Body of Detroit (which later established a local manufacturing plant under the name Murray-Ohio, which became a major bicycle manufacturer); seats: 26; front and rear doors; unnumbered 26-passenger chain-driven "Opera Bus," operated for a short time in 1904 by CER; front and rear doors. The name "opera" is believed to be from the bus's use in loop service to the Cleveland Opera House.

SECTION I
## Buses Purchased by the Cleveland Railway Company

### Fleet Statistics: White, 50–52

BLDR: White 1914; body: G. C. Kuhlman; seats: 26; body type M; front door chain driven; disposition unknown.

Buses 50–52, experimental fleet of Cleveland Railway buses, White-Kuhlman, 1914. *(Barbara Branch)*

### Fleet Statistics: Hood-in-Front, 100–129

BLDR: G. C. Kuhlman; components and construction: White; LEN: 28' 10"; seats: 29; model: 50A; W: 12,900 lbs; engine: White 4 CYL gasoline; capacity fuel tank: 34 gals; front doors. All buses delivered between 8-11-25 and 1-23-26; all sold for SCR 10-15-38.

123  dismantled after ACC hit by train, SCR 12-16-30 (except chassis, which was used to construct bus 170 in 1929)

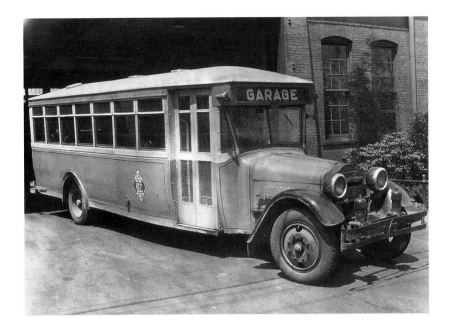

Bus 127, first regular fleet of Cleveland Railway buses, White-Kuhlman, 1925. *(Cleveland Railway photo, Blaine Hays Collection)*

### Fleet Statistics: Hood-in-Front Double Deck, 500–558

BLDR: Safeway Six Wheel, Philadelphia; LEN: 33'; seats: 31 lower, 31 upper; two-man, driver and conductor (front entrance, rear exit: conductor stationed at rear platform and oversaw treadle-controlled rear door while driver had manual-operated front door); model: 66; engine: Continental 6 Cyl gasoline; capacity fuel tank: 38 gals; front and rear doors; 500–529 D 1925; 530–549 D 1926; 550–552 D 1927; 553–558 D 1928; first day of double-deck service 11-7-25; in only rush-hour service after 1931; all OOS 1-31-36. Some double-deck frames became work units as follows:

| | |
|---|---|
| 509 | W 19,300 lbs, cut down to single decker 1931 as a test to determine if fleet could be salvaged, not successful, sold for SCR 10-15-38 |
| 530 | retained for work service |
| 531 | REB to salt spreader and REN A-13 on 12-3-37, assigned to Windermere |
| 533–539 | retained for work service |
| 543 | REB to salt spreader and REN A-14 in 12-1938, assigned to Superior |
| 544 | REB to salt spreader and REN A-15 in 12-1938, assigned to Windermere |
| 546 | retained for work service |
| 547 | retained for work service |
| 556 | the oddball double decker, had restyled body paneling, a larger destination sign, and two rub rails instead of one; was sold to CTS General Manager Walter J. McCarter and moved to East Clarion, OH, CON to cottage, later SCR |

Double-deck bus 518, six-wheel, 1925. *(Cleveland Railway photo, Blaine Hays Collection)*

Bus 509, CON to single-deck, 1931, six-wheel, 1925. *(Cleveland Railway photo, Blaine Hays Collection)*

### *Fleet Statistics: Versare One-of-a-Kind, 700–701*

BLDR: Versare Corp.; body: Versare, Albany, NY; seats: 37; engine: Waukesha

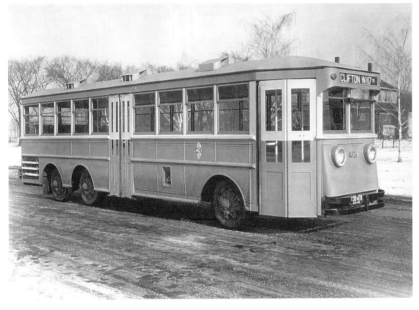

Bus 700, FN 400, "The Monster," Versare, 1926. *(Cleveland Railway photo, Blaine Hays Collection)*

Bus 401, REN 701, Versare, 1928, gas-electric. *(Cleveland Railway photo, Blaine Hays Collection)*

700    D 9-11-26, FN 400, hood-in-front; W 24,540; LEN 38' 10"; 8-wheel gas-electric model, capacity fuel tank 52 gals; had railroad roof with windows and a smoking compartment, electric drive transmission by WH, 6 CYL gasoline; front doors on either side and a right side center door, BLT for Alton Transportation Co. for service between Jacksonville and East

St. Louis, IL, ACQ used by Cleveland where it came to be known by employees as the "Monster," sold for SCR 10-15-38

701    D 1-18-28, FN 401; W 20,200; LEN 31' 2"; 6-wheel gas-electric model; capacity fuel tank 60 gals, 6 CYL gasoline; first non-hood-in-front bus in Cleveland; front and center doors; electric drive transmission by WH; built for Alton Transportation Co., for service between Jacksonville and East St. Louis, IL, ACQ used by Cleveland, sold for SCR 10-15-38

### Fleet Statistics: Hood-in-Front, 130–159

BLDR: G. C. Kuhlman; components and construction: White; LEN: 28' 10"; seats: 29; model: 50A; W: 12,900 lbs; engine: White 4 CYL gasoline; capacity fuel tank: 34 gals; front doors; original series consisted of 130–153 with an add-on order three months later of 154–159 with a change to model 50B. All D between 2-6-26 and 9-27-26; all sold for SCR 10-15-38.

Bus 149, hood-in-front type, White, 1926. *(Cleveland Railway photo, Blaine Hays Collection)*

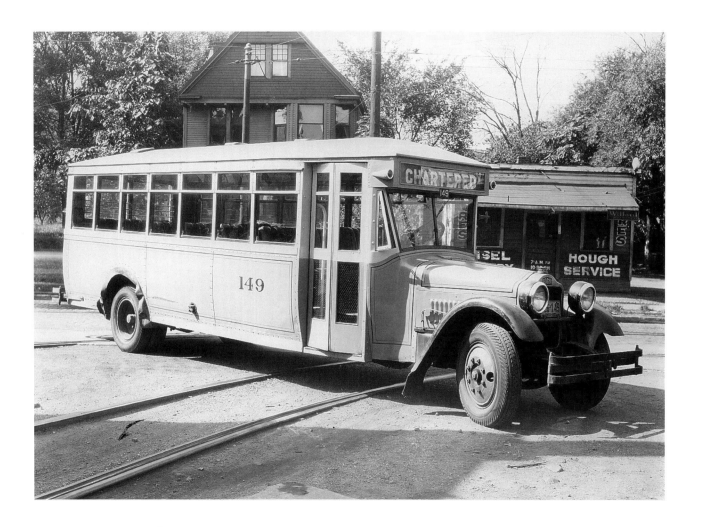

152 After its retirement in 1938 sat along with several others at Harvard Shops. ACQ by Way & Structures employee William Rae and moved to Unionville in Ashtabula County in 1942 where front was torn off and passenger compartment CON to a shed. In 1988 discovered by bus historian Mark Urban of Lakewood and offered to RTA for preservation. RTA accepted the donation, purchased accurate engine to go with it, and moved bus body to Central Bus Maintenance 3-1989. RTA experienced a management change in its maintenance division and project of restoring the bus fell from favor. Body was SCR 1993.

### Fleet Statistics: Hood-in-Front, 300

BLDR: 6-Wheel Co. (Safeway), Philadelphia; LEN: 28' 10"; W: 16,060 lbs; seats: 29; model: 64; engine: Continental 6 CYL gasoline; capacity fuel tank: 35 gals; front and rear doors; D 12-3-26; Timken bearings; OOS 1920s; sold for SCR 10-15-38.

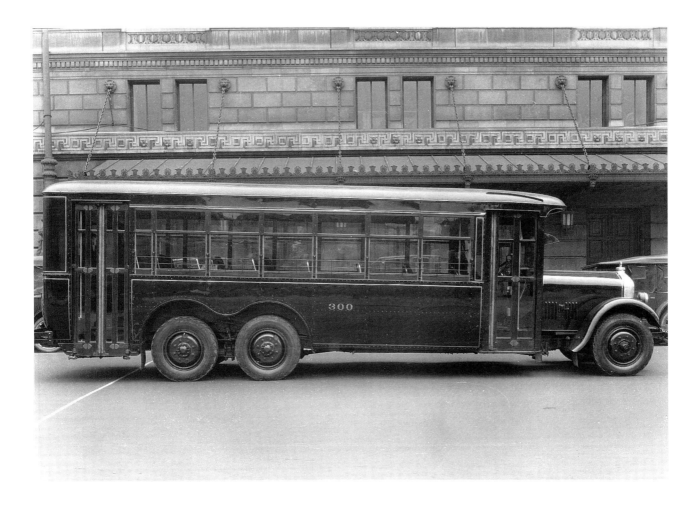

Bus 300, hood-in-front type, Safeway six-wheel coach, 1926. *(Cleveland Railway photo, Blaine Hays Collection)*

### Fleet Statistics: Hood-in-Front, 301–310

BLDR: Yellow Coach; body: Lang Body; LEN: 29' 6"; seats: 29; model: Z-230; W 16,200 lbs; engine: Continental 6 CYL gasoline; capacity fuel tank: 55 gals; front doors; ACQ for Heights express service; D between 12-27-27 and 1-10-28; all sold to Apex Steel & Supply 6-1940.

### Fleet Statistics: Hood-in-Front, 160–170

BLDR: White; body: Lang Body; LEN: 30'; seats: 29; model 50B; W: 14,900 lbs; engine: White 4 CYL gasoline; capacity fuel tank: 55 gals; front and rear doors. All D 10- to 12-1927.

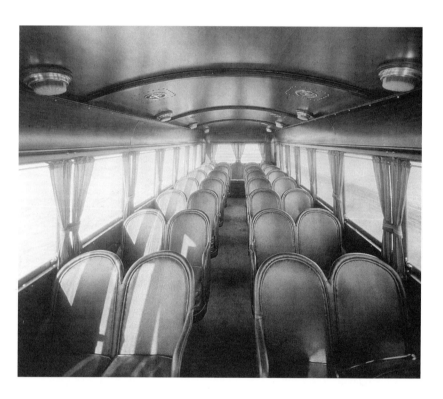

Cleveland's early hood-in-front buses featured individual bucket seats, window curtains, and narrow cramped aisles. (*Cleveland Railway photo, Blaine Hays Collection*)

160–165  body SCR 1938, and chassis sold to Bassicus & Sons 7-12-40
166–169  sold for SCR 10-15-38
170      assembled by CR using chassis of bus 123, SCR 10-27-39

### Fleet Statistics: Hood-in-Front, 200–219

BLDR: White; body: Lang Body; LEN: 30' 4"; seats: 29; model: 54; W: 15,700 lbs; engine: 6 CYL gasoline; capacity fuel tank: 55 gals; front and rear doors. All D between 1-20-28 and 1-31-28.

201, 202, 204, 209, 212, 213, 216–219 • sold Bassicus 7-1940

200, 203, 205–208 210, 211, 214, 215 • SCR 1940, but chassis retained

### Fleet Statistics: Hood-in-Front, 400–429

BLDR: White; body: Lang Body; LEN: 32' 7"; seats: 38; model: 54A; W: 18,360 lbs; engine: White 6-CYL gasoline (421–429, Continental 6 CYL); capacity fuel tank: 65 gals; front and rear doors. ACQ to replace Scranton Avenue streetcars. One bus from this series, believed to be 429, was CON to work unit. All D between 4-1-29 and 4-29-29.

400–419 sold to Bassicus & Sons 7-1940

420, 422, 423, 426, 427, 428 • bodies assembled on used chassis ACQ by CR from White

421, 424, 425 • sold 6-26-41 to Belle St. Louis

429 assembled by CR, storage 1941, CON to Properties Department bus in 1947 and REN B-5, reassigned to Equipment

Bus 429, hood-in-front type, White, 1929. *(Cleveland Railway photo, Blaine Hays Collection)*

Department 1949 and stored at Berea Garage, reassigned to Buildings Department 1950 and moved to Harvard Shop, OOS 1952, operated in "Parade of Progress" 4-27-52, stored at Denison Station filled with junk. All other "B"-series vehicles were trucks except B-13, which appears in the 220–274 series, Fleet 17

### Fleet Statistics: Hood-in-Front, 600–604

BLDR: Yellow Coach; body: Lang Body; LEN: 31'; seats: 39; model: Z-240; W: 17,900 lbs; engine: Continental 6 CYL; capacity fuel tank: 65 gals; front and rear doors; all D between 3-11-29 and 3-16-29; all sold 6-1940 to Apex Steel & Supply.

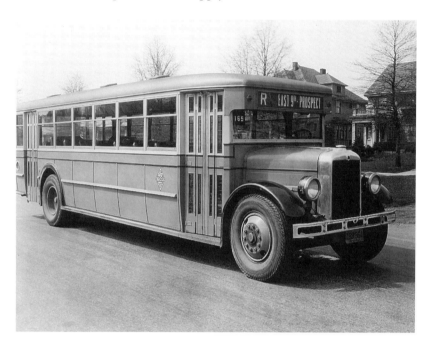

Bus 603, hood-in-front type, Yellow Coach, 1929. *(Cleveland Railway photo, Blaine Hays Collection)*

### Fleet Statistics: Hood-in-Front, (800) 801–810

BLDR: Indiana Coach; body: Bender Body; LEN: 23' 3"; seats: 17; model: 14-B; W: 9,300 lbs; engine: Hercules J.X.C. 6 CYL gasoline; capacity fuel tank: 35 gals; front door; all D between 3-11-33 and 4-5-33; 803 and 807 sold to Shepherd Bus Co. 1940.

Note: Coach 800 was placed in this fleet for convenience in numbering.

800   FN 600, BLT Twin Coach, LEN 31'10", Seats 40, Model: Urban; Engine: Hercules W.X.E. (used), 6 CYL gasoline, W 17,350 lbs, front and rear doors, sold 7-18-40 to Bassicus & Sons

Bus 600, REN 800, Twin Coach, 1927. *(Cleveland Railway photo, Blaine Hays Collection)*

| | |
|---|---|
| 801 | storage 8-22-38, CON to service bus for Sewage Department |
| 802 | OOS 7-5-38, storage 7-5-38, chassis only CON to E. 55 Station training school |
| 804 | OOS 7-5-38, storage 7-5-38, sold |
| 805 | OOS 7-5-38, storage 7-5-38, sold |
| 806 | OOS 8-23-38, storage 8-23-38, sold |
| 808 | OOS 6-23-38, storage 6-23-38, sold |
| 809 | storage 9-14-38, CON to service bus for Track Department |
| 810 | OOS 6-23-38, storage 6-23-38, sold |

Bus 808, hood-in-front type, Indiana Coach, 1933. *(Cleveland Railway photo, Blaine Hays Collection)*

### Fleet Statistics: White, 710–775

BLDR: White; body: Bender Pancake (flat MOT); LEN: 28' 3"; seats: 32; model: 684; W: 16,400 lbs; engine: 12 CYL White 10-A gasoline; capacity fuel tank: 67 gals; front and center doors. Bus 710 D 1934; 711–730 D 1935; 731–775 D 1936; 710–730 ACQ to temporarily replace Payne Avenue streetcars prior to inauguration of TT service; 731–775 replaced streetcars on Central Avenue. All sold for SCR 1946–48.

710    delivered painted orange, after use as demo for another system
726    had air clutch

Bus 740, White, 1936. *(Cleveland Railway photo, Blaine Hays Collection)*

### Fleet Statistics: Ford, 450–474

BLDR: Ford, Union City; LEN: 29' 7"; seats: 25; model: 81-B; W: 10,320 lbs; engine: Ford BB-15 V-8 gasoline; capacity fuel tank: 45 gals; front and center doors. Poor performance vehicles, were replaced by Ford with 840–899 series.

Note: Fleets from 450 to 574 were ACQ from five manufacturers to avert criticism that CR favored local White Co. All D 5-13-38 to 6-18-38; all sold 1940–41.

459, 460, 470, 471 • sold to Latimen Co. 2-1941

### Fleet Statistics: Twin, 475–499

BLDR: Twin Coach; LEN: 26' 6"; seats: 27; model: 27-R; W: 10,700 lbs; engine: Hercules WXC-3 6 CYL gasoline; capacity fuel tank: 65 gals; front and center doors; all D between 4-12 and 4-15-38; all OOS 1948 and sold for SCR 1948.

### Fleet Statistics: Mack, 500–524

BLDR: Mack; seats: 31; model: CQ; engine: 6 CYL gasoline; front and center doors; all D 1938; OOS 1947–48; sold for SCR 1949–50.

### Fleet Statistics: Yellow, 525–549

BLDR: Yellow Coach; seats: 31; model: 728; engine: gasoline; front and center doors; all D 1938; all sold for SCR 1949–50.

### Fleet Statistics: White, 550–574, Fleet 11

BLDR: White; LEN: 28' 1"; seats: 31; model: 784; W: 15,380 lbs; engine: 12 CYL gasoline; capacity fuel tank 85 gals; front and center doors; all D 1938; all sold for SCR 1952–54. Used to pull trailers during World War II.

Bus 550, White, 1938. *(White Company photo, Blaine Hays Collection)*

550, 554, 560, 566, 567 · sold 6-7-54 to Cherry Bros.

554    had experimental diesel engine, first in Cleveland

555    CON to Safety Coach for Transportation Department and REN Z-99, sold 5-25-59

565    operated in "Parade of Progress" as "Tribe Ticket Trolley" 4-27-52

567    operated in "Parade of Progress" pulling trailer T-97 4-27-52

### Fleet Statistics: Twin, 605–614

BLDR: Twin Coach; seats: 27; model: 27-R; engine: gasoline; front and center doors; all D 1939; OOS 1947; sold for SCR 1948.

Bus 612, Twin Coach, 1939. *(Cleveland Railway photo, Blaine Hays Collection)*

### Fleet Statistics: White, 645–674

BLDR: White; LEN: 33'; seats: 40; model: 788; W: 16,950 lbs; engine: 12 CYL gasoline; capacity fuel tank: 105 gals; front and center doors; 645–660 D 1939, OOS 1950, sold for SCR 1954; 661–674 D 1940; ACQ to replace Wade Park streetcar service; sold for SCR 1955–57. Used to pull trailers during World War II.

646    CON to Power & Plant material storage bus 1956, REN P-646

657    CON to Power & Plant material storage bus, sold 3-20-57

659    CON to Wheel Balancing Service bus and REN A-97 on 12-28-55, OOS 4-24-57, sold 6-5-57

663    D 1940, CON to Denison Station Transportation Department magazine coach, sold 3-20-57

664   D 1940, caught on fire, SCR 1947
666   D 1940, CON to Power & Plant material storage bus, sold 2-3-58

### Fleet Statistics: Yellow, 685–694

BLDR: Yellow Coach; LEN: 33'; seats: 40; model: TD-740; W: 18,900 lbs; engine: 6 CYL diesel; capacity fuel tank: 85 gals; front and center doors; first fleet of diesel buses ACQ for Cleveland; D 1940; all sold to scrapper L. H. Samworth April 1957 except for:

685   caught fire, SCR 1947
686   CON to Safety Coach for Transportation Department and REN Z-98, sold to Babcock Co. 5-25-60

### Fleet Statistics: Mack, 625–634, Fleet 16

BLDR: Mack; seats: 26; model: CY; engine: gasoline; front and center doors; all D 1940; OOS 1952; sold for SCR 1953.

627   CON to St. Clair Station farebox storage coach, SCR 12-28-53
628   CON to Denison Station farebox storage coach, SCR 12-28-53

### Fleet Statistics: Yellow, 350–369, Fleet 20

BLDR: Yellow Coach; LEN: 33'; seats: 40; W: 16,800 lbs; engine: 6 CYL diesel; model: TD (Transit Diesel); capacity fuel tank: 85 gals;

Bus 820, FN 350, Yellow Coach, 1940. *(Cleveland Railway photo, Blaine Hays Collection)*

front and center doors; 350–354 D 1940 (FN 820–824 REN 1941), model TD 4000; 355–369 D 1941, model TD 4001; all SCR 1955–56.

359, 360, 362, 364, 368 • sold 12-1955 to Ohio Valley Bus

### Fleet Statistics: White, 220–274, Fleet 17

BLDR: White; LEN: 32' 11 15/16"; seats: 40; model: 788; W: 17,400 lbs; engine: 12 CYL gasoline; capacity fuel tank: 105 gals; front and center doors; 220–234 D 1940 (FN 825–839, 6-1941), all SCR by 1956; 235–274 D 1941, ACQ to replace streetcars on E. 30th and E. 79th streets. Used to pull trailers during World War II.

Bus 226, REN to work bus A-96, White, 1940. *(Cleveland Transit photo, Blaine Hays Collection)*

| | |
|---|---|
| 220 | FN 825 (6-4-41), CON CTS gardener coach and REN B-13 on 12-5-55, sold 3-20-57 |
| 224 | FN 829 (6-3-41) CON to Power & Plant storage bus and REN P-98 on 3-26-56, sold 1-28-57 |
| 226 | FN 831 (6-4-41) CON to Stores Department tire service bus, REN A-96, later REN M-96, SCR 1956 |
| 230 | FN 835 (6-4-41) CON to Power & Plant storage bus, REN P-97 on 10-17-56, sold 6-5-57 |
| 232 | FN 837 (6-3-41) CON to storeroom at Fairmount Station, sold 8-28-58 |
| 243 | placed in storage for Plant Department 8-6-54 |
| 244 | sold to Murray Excavation 8-26-54 |

247 caught fire, used for Power & Plant tool storage, SCR 1954
252 stored for Stores Department on 8-6-54
257 sold to City of Cleveland, 1955
263 sold to Aurora Construction in 1954

### Fleet Statistics: Mack, 275–299, Fleet 23

BLDR: Mack; LEN: 29' 5"; seats: 32; model: RC; W: 14,400 lbs.; engine: 6 CYL gasoline; capacity fuel tank: 65 gals; front and center door; all D 1941; SCR 1953–54.

Bus 292, Mack, 1941. *(Cleveland Transit photo, Blaine Hays Collection)*

275, 281, 282, 283, 287, 289, 291, 297 • sold to Cherry Bros. Iron & Steel 5-1954

### Fleet Statistics: Mack, 311–335, Fleet 19

BLDR: Mack; LEN: 30'; seats: 40; model: CM; W: 18,800 lbs; engine: 6 CYL gasoline; capacity fuel tank: 80 gals; front and center doors; 311–315 D 1940 (FN 815–819); 316–335 D 1941; all SCR 1953–54.

318–320, 324, 325, 328, 330, 332, 335 • sold Cherry Bros. 1954

### Fleet Statistics: Mercury, 840–899

BLDR: Ford, Mercury Division; seats: 27; model: 09B engine, gasoline; front and center door; 840–849 D 1941; 850–899 D 1940. All sold

Bus 316, Mack, 1941. *(Cleveland Transit photo, Blaine Hays Collection)*

Bus 845, Mercury, 1941. *(Cleveland Transit photo, Blaine Hays Collection)*

224 · CLEVELAND'S TRANSIT VEHICLES

for SCR 9-30-52 except 850 and 867 sold 11-29-49 and 886 and 891 sold for SCR 12-28-53.

### Fleet Statistics: Twin, 180–189

BLDR: Twin Coach; model: 30GW; ACQ 1942; extra wide, and joined by purchase of other wide buses; D between 2-19-42 and 2-26-42; poorly constructed because of wartime production limitations; all OOS 1947; size made them unfit for application in Cleveland; sold for SCR 1948.

## SECTION 2
## Buses Purchased New by the Cleveland Transit System

### Fleet Statistics: Bus Trailers, TC100–TC160

BLDR: Schult Trailers; D between 11-1942 and 11-1943. All were placed into service upon arrival. These plywood trailers were constructed in a hurry for CTS on old auto-hauler chassis. They arrived without seats but later underwent extensive modifications before re-entering service. Most assigned to Superior Station, but one at Brooklyn Station. They were used on Wade Park, E. 79th, and other lighter lines. Trailers discontinued after the war, and certain ones REN into the miscellaneous service equipment fleet and continued in use. They were:

T-90   SCR 1-21-53
T-91   SCR 1-13-54

Schult bus trailer, 1943, pulled by White 562. *(Cleveland Transit photo, Blaine Hays Collection)*

T-92  SCR 1-21-53
T-93  SCR ?
T-94  SCR 6-11-54
T-95  sold 7-14-53
T-96  SCR ?
T-97  operated in "Parade of Progress" 4-27-52, SCR 5-23-56
T-98  SCR ?
T-99  later REN S-99 when used by transportation department, SCR 1-26-53

Note: The others were sold mostly to veterans for use as cottages directly after the war.

### Fleet Statistics: White, 3000–3024, Fleet 30

BLDR: White; LEN: 32' 11$^{15}/_{16}$"; seats: 40; model: 788; W: 18,500 lbs; engine: 12 CYL gasoline; capacity fuel tank: 103 gals; stick-shift coaches; D 1945; OOS 1954–56 and sold for SCR.

3021  held for use by Power & Plant Department, OOS 12-28-55
3024  CON to Power and Plant coach and REN P-99, OOS 12-28-55, sold 1958

### Fleet Statistics: Yellow-GM, 3200–3224, Fleet 32

BLDR: Yellow Coach/GM; LEN: 33'; seats: 40; model: TDH-4007; W: 17,000 lbs; engine: 6 CYL diesel; capacity fuel tank: 85 gals; front and center doors; D 1945; sold as follows:

Bus 3200, Yellow-GMC, 1945. *(Cleveland Transit photo, Blaine Hays Collection)*

3200–3201, 3203, 3205, 3208–3209, 3213–3215, 3218, 3222 • Des Moines, 1960
3202, 3221 • Ohio Valley Bus, 1955
3204, 3210, 3212, 3220 • Utica Transit, 1960
3206–3207, 3211 • Warren Transit, 1960
3216–3217   North Olmsted Bus Line, 1960
3219, 3223–3224 • Transportation Assn., 1960

### Fleet Statistics: White, 3800–3854, Fleet 06

BLDR: White; LEN: 34' 11 15/16"; seats: 44; model: 798; W: 18,500 lbs; engine: 12 CYL gasoline; capacity fuel tank: 130 gals; front and center doors; gear-shift type; ACQ 1946; OOS 1955–57; sold as follows:

| | |
|---|---|
| 3800 | City of Maple Heights, 1956 |
| 3810, 3839, 3844 • Kenny 1957 | |
| 3812 | CON Power & Plant work bus, REN P-147, sold 1958 |
| 3813 | CON Power & Plant work bus, REN P-148 on 5-27-57, sold 1958 |
| 3822 | Broadview Bus Lines 1956 |
| 3829 | Fehlner 1955 |
| 3838 | City of Cleveland, Department of Public Safety and CON to Cleveland Police riot control bus 4-25-57 |
| 3845 | CON to Power & Plant work bus, REN P-146 12-19-56, sold 1957 |
| 3851 | CON to Bert Kenehan's emergency bus, REN E-98 4-29-57, sold 1958 |

Note: This fleet purchased under first-series equipment trust dated 6-1-46.

### Fleet Statistics: Super-Twin Articulated

BLDR: Twin Coach 1946, operated in Cleveland for a short time in 1946 in demonstration service.

### Fleet Statistics: Twin, 3648–3659, Fleet 8

BLDR: Twin Coach; LEN: 34' 10"; seats: 44; model: 44D; W: 15,400 lbs; engine: 6 CYL gasoline; capacity fuel tank: 80 gals; front and center doors. D as dual-engine buses but later CON to single engine, but never operated properly thereafter since out of balance. D 1946; 3650–3659 OOS 1958–60. Sold as follows:

3648, 3649 • FN 304-305 of Lakewood Rapid Transit, model 34S, REB but never used, OOS 12-28-55,
3650   LaPine Truck Sales 1960
3651, 3653, 3657 • Ours 1960
3652   St. Michael's Church 1960
3655   Machaud Baking 1959
3656   City Auto Wrecking 1958
3658   St. Mary's Church 1959
3659   John Adams High School 1959

Note: 3650–3659 purchased under first-series equipment trust 6-1-46.

Super-Twin articulated bus demonstrator, 1946. *(Cleveland Transit photo, Blaine Hays Collection)*

### Fleet Statistics: GM, 3395–3425, Fleet 12

BLDR: GM; LEN: 35'; seats: 44; model: TDH-4507; W: 16,330 lbs; engine: 6 CYL diesel; capacity fuel tank: 85 gals; front and center doors; 3395–3399 (FN 3400–3404) D 1946, remainder 7-1947 to 8-1947. Sold as follows:

3395–3599   Dayton 1959
3400   Dayton-Western 1960
3401, 3424 • Geauga Transit 1964
3402, 3411 • Lorain City Lines 1964
3404, 3407, 3418, 3420 • St. John's Trans. 1964
3405–3406, 3408, 3412–3414, 3416, 3419, 3422–3423 • Memphis Transit 1960
3415   operated to tank plant as "Bomber Express," Kricfalusy 1964
3417   ACC 10-23-56, body parts salvaged, SCR 4-3-57

Bus 3400, GMC, 1947. *(Cleveland Transit photo, Blaine Hays Collection)*

3421    ACC 10-15-58, City Auto Wrecking

3425    (Fleet 23) D 10-1953 as demonstrator, numbered 3425 when ACQ by CTS to provide sufficient buses to replace Madison streetcars, REN C-350 (Fleet 26) 3-7-57 although "C" was never physically applied to coach, sold Hausman Bus Sales 1969

Carrying large numbers of workers to the Brook Park Tank Plant was this old-style GM bus nicknamed the Bomber Express. *(Cleveland Transit photo, Blaine Hays Collection)*

PART D: MOTOR COACH ROSTER · 229

Note: 3400–3424 purchased under first-series equipment trust dated 6-1-46.

### Fleet Statistics: Twin, 3660–3709, Fleet 21

BLDR: Twin Coach; LEN: 34' 10"; seats: 44; model: 44S; W: 15,400 lbs; engine: 6 CYL gasoline; capacity fuel tank: 80 gals; front and center doors; D 9-1947 to 10-1947; OOS 1962–64. Sold as follows:

3660, 3664–3665, 3672, 3674, 3678–3679, 3685 • Cleveland Southeastern 1964
3661, 3668, 3670, 3700, 3701 • Tee Pee Motors 1964
3662    Moon Glow Skating Rink 1964
3663    Fairhill Hospital 1964
3666    Tee Pee Motors 1964; preserved at Ohio Museum of Transportation, Orrville
3671    Smith 1964
3673, 3675, 3680, 3682, 3686, 3689, 3691, 3696, 3707–3709 • Kricfalusy, 1964
3677    Kubentz 1964
3681    St. Stephen's 1963
3687, 3690, 3692, 3697 • Rochester, N.Y., 1963
3693    Maresch 1964
3694    Maurice Young 1964
3695, 3699 • Chagrin Falls Skating Rink 1963
3698    Ours 1960
3702    Fire, SCR
3703    CON tow bus at Woodhill Station 3-29-61, OOS 4-26-62, sold 1962
3704    Western Reserve Transit, Youngstown, 1962
3705    Babcox 1960

Note: This fleet purchased under second-series equipment trust dated 12-15-46.

### Fleet Statistics: Twin 3710–3784 (3785), Fleet 15

BLDR: Twin Coach; LEN: 32' 11"; seats: 41; model: 41S; W: 13,500 lbs; engine: 6 CYL gasoline; capacity fuel tank: 80 gals; front and center doors; 7-1947 and 11-1947; OOS 1959–63. Sold as follows:

3711, 3714, 3721, 3729 • North Olmsted Municipal Bus Lines 1962
3712, 3742, 3770 • Kubentz 1964
3713, 3777 • Babcox 1960

Bus 3683, Twin, 1947. (Cleveland Transit photo, Blaine Hays Collection)

3715     OOS 4-13-60, held for conversion to Power & Plant work bus P-149 but not used, sold

3716     used for tow bus at Triskett Station, Kricfalusy 1960

3717, 3719, 3722, 3730, 3735, 3758, 3769, 3776, 3780 • Kricfalusy 1960

3718     CON to Power & Plant workshop and REN P-150, 1-17-64

3723, 3759, 3768, 3778 • St. Stephen's, Niles, OH, 1963

3724     St. Helen's Church, Newbury, OH, 1963

3727, 3745, 3752, 3779 • Cleveland Southeastern Trails 1962

3733     Cleveland Construction, 1964

3734     used as tow and service bus at Woodhill Station, OOS 4-26-62, sold 1962

3743, 3746, 3749–3750, 3753, 3756, 3760–3762, 3764–3765, 3771–3772 • Dayton 1959

3744     in serious front-end ACC, stored Superior, SCR

3751     Kirkham Service, 1964

3754     ACC 6-21-56, SCR 7-27-56

3766     Hawthorn 1962

3773     used as tow bus at Hayden Station, OOS 4-26-62, sold 1962

3775     City Auto Wrecking 1959

3781, 3782 • Tee Pee Motors 1964, 3781 preserved at Ohio Museum of Transportation, Orrville

3783     Terra Vista 1964

3784 Dana Corp. 1964

3785 D 1951, propane fueled, REN 2400 in 1954 and lavishly outfitted as "The Charterchief," first ran as regular two-door bus; operated in "Parade of Progress" 4-27-52; center door sealed when CON to Charterchief. As it was the only propane-fueled bus in Cleveland, a special fueling station was located at Windermere Station to service this one coach; in 1954 the Equipment Department tired of this operation and CON bus to gasoline. In 1961 REN 2400 and PTD a light-blue with four different colored diamond shapes (in keeping with diamond logo of the Motor Coach Department) on the side and outfitted as System's Charterchief at $10,000 cost; sold to Howard Enterprises 5-8-69 and spent some time in Akron area; returned to Cleveland and resides on grounds of Doc's Auto Clinic at 2948 E. 81st near Kinsman and still sports its four diamond targets and has "Eureka" above the front door.

Note: 3710–3734 purchased under first-series equipment trust 6-1-46; 3735–3784 purchased under second-series equipment trust 12-15-46.

### Fleet Statistics: White, 3855–3954, Fleet 13

BLDR: White; LEN: 34' 11$^{15}/_{16}$"; seats: 44; model: 798; W: 19,300 lbs; engine: 12 CYL gasoline; capacity fuel tank: 130 gals; front and center doors; hydro-torque automatic shift; ACQ 1947; OOS 1957–59. Sold as follows:

3855, 3858, 3863–3866, 3870, 3872–3874, 3876–3877, 3880, 3882–3883, 3887–3888, 3895–3899, 3906–3916, 3919, 3922–3923, 3932, 3941–3943, 3951, 3953 • J. Ours 1959

3857, 3861, 3871, 3885, 3900, 3907–3908, 3913, 3915, 3925, 3927, 3935, 3937, 3940, 3954 • Kubentz 1960

3859 Holy Name School

3868 CON to Power & Plant workshop REN P-149, 5-29-58, sold Kubentz

3869, 3881, 3893–3894, 3948 • City Auto Wrecking 1959

3878 OOS 3-25-57 ACC, sold 1957

3889–3990, 3903 • City of Maple Heights 1960

3891, 3911, 3933 • Kricfalusy 1959

3892 Twin Coach Co.

3902 CON to British Leyland Diesel engine, sold Berea Bus Lines 1959

3905 St. Ignatius High School 1959

3945 severe fire damage, SCR

3949 ACC front end, sold 1-28-57

Bus 3868, White, 1947. *(Cleveland Transit photo, Blaine Hays Collection)*

Note: This fleet purchased under second-series equipment trust 12-15-46.

### Fleet Statistics: Transit, 2600–2603, Fleet 26

BLDR: Transit Bus Co.; LEN: 29' 4"; seats: 32; model: 81; W: 13,980 lbs; engine: Ford 6 CYL diesel; capacity fuel tank: 90 gals; front and center doors. Coaches ACQ by South Euclid 7-1948 for a new South Euclid Express service assumed later by CTS. Coaches ACQ by CTS and assigned to Fairmount Station; used on Fairmount shuttle to replace streetcars. Sold as follows:

Bus 2600, Transit bus, 1948. *(Cleveland Transit photo, Blaine Hays Collection)*

2600, 2602, 2603 • Lorain Bus Lines 1953
2601 Transit Sales Bridgeport 1953

### Fleet Statistics: GM, 3405–3424

BLDR: GM; these were new buses delivered to Louisville Railways Co. in trade for PCC cars. REN in sequence in Louisville to same numbers the PCC cars had received, 501–520.

### Fleet Statistics: GM, 3450–3475, Fleet 12

BLDR: GM; LEN: 35'; seats: 44; model: TDH-4507; W: 16,330 lbs; engine: 6 CYL diesel; capacity fuel tank: 85 gals; front and center doors; D 1-1948. Sold as follows:

3450, 3452 • Geauga Transit 1964
3451, 3453, 3454, 3456, 3459, 3462, 3465, 3467–3471, 3474 • St. John's Trans. 1964
3453 Dayton-Western Motors 1964
3455, 3457, 3458, 3472 • Lorain City Lines 1965
3463 CON Safety Department coach and REN Z-99 1964
3466 in serious front-end ACC
3473 held for conversion to safety bus, sold St. John's Trans. 1965
3475 former Greyhound, D 6-1961 from Broadview Bus Lines (FN 7), sold Lorain City Lines 1964

Note: 3450–3474 purchased under second-series equipment trust 12-15-46.

### Fleet Statistics: GM, 3150–3199, Fleet 25

BLDR: GM; LEN: 30' 9"; seats: 36; model: TDH-3610; W: 14,800 lbs; engine: 4 CYL diesel; capacity fuel tank: 85 gals; front and center doors; ACQ 1948. Sold as follows:

3150, 3151, 3160, 3161, 3164, 3177, 3183 • Philadelpia Transit 1956
3184 in serious ACC 1949
3198, 3199, 3155–3158, 3162, 3168, 3178, 3182, 3185–3188, 3190, 3191, 3194, 3196 • Dayton 1959
3153, 3159, 3165–3166, 3169, 3172, 3174–3175, 3179–3180 • Ohio Rapid (Columbus) 1956

3154, 3167, 3170, 3171, 3173, 3176, 3181, 3189, 3192–3193, 3195, 3197 • Delaware Coach 1956

3163   Lake Shore Coach

Note: This fleet purchased under second-series equipment trust 12-15-46.

### Fleet Statistics: GM, 3110–3149, Fleet 27

BLDR: GM; LEN: 30' 9"; seats: 36; model: TDH-3612; W: 14,800 lbs; engine: 4 CYL diesel; capacity fuel tank: 85 gals; front and center doors; ACQ 1949. A popular model, these were sold for a price CTS could not afford to pass up. Replaced by 700–739. Sold as follows:

3110, 3111, 3117, 3123, 3126, 3128, 3130, 3133, 3136, 3137, 3139, 3146–3149 • sold Banco de Fomento Commercial, Havana, Cuba 1955
3112   L. Bannon 1955
3113, 3115 • Mr. Grasso, Union City, NJ, 1954
3114, 3129 • Dayton-Xenia Lines
3116, 3125, 3127, 3134, 31389, 3142, 3145 • Danville (VA) Power & Traction 1955
3124, 3141, 3143 • Price & Co. 1954
3131, 3132, 3135 • Lafayette Greenville Lines 1955
3140, 3144 • LaCrosse, WI, 1955

Note: 3110–3126 purchased under second-series equipment trust 12-15-46.

### Fleet Statistics: GM, 3500–3631, Fleet 35

BLDR: GM; LEN: 39' 9"; seats: 51; model: TDH-5103; W: 19,200 lbs; engine: 6 CYL diesel; capacity fuel tank: 85 gals; front and center doors; D 1950–52; OOS 1970–71. Sold as follows:

3500–3501, 3504–3505, 3508, 3511–3514 • Lake Front Lines 1970
3502–3503, 3506–3507, 3509–3510, 3515–3516, 3519–3521, 3523–3526, 3529–3531, 3533, 3539, 3541, 3543, 3546, 3557, 3568–3569, 3572, 3574, 3578, 3582, 3584–3585, 3588–3590, 3599, 3605–3506, 3608, 3610–3511, 3613–3516, 3619–3520, 3628 • Dayton-Western Motors 1971
3517, 3522, 3527, 3532, 3534–3536, 3538, 3540, 3549, 3551, 3552, 3555, 3561–3564, 3566, 3576–3577, 3580, 3586–3587, 3595, 3602, 3604, 3607, 3609, 3617, 3621, 3623 • Ours 1971

3518    J. Carithers 1971
3528, 3537, 3544, 3558, 3565, 3592–3594, 3596, 3598, 3600–3601, 3612, 3622, 3624–3625, 3627 • Del-Val Coach 1971
3542, 3603, 3626 • E. Wheatly 1970
3545, 3547, 3550, 3553, 3559 • Donald Hines 1971
3548, 3554, 3560, 3567, 3591, 3597, 3618 • Maple Heights Transit 1970
3556    C. Gohr 1970
3570, 3571 • Dean's Truck Body 1971
3573, 3575, 3579 • E. Snider 1971
3581, 3583 • B. Dyer 1971
3594    CON to Zoo bus during 1950s
3629    operated in "Parade of Progress" 4-27-52, J. Oblak 1969
3630    D 3-17-52, demonstrator, OOS 6-1970, Ours 1970
3631    D 1953, demonstrator, model GM-TDH 5105, later REN 525, Maple Heights Transit 1971

### Fleet Statistics: Mack, 2700–2749, Fleet 27 (Later Fleet 28)

BLDR: Mack; LEN: 39' 11"; seats: 50; model: C 50 DT; W: 21,765 lbs; engine: 6 CYL diesel; capacity fuel tank: 85 gals; front and center doors. Shortly after D of this fleet in 6-1951 and 7-1951, an enormous amount of trouble developed with the Cummins Diesel engines in these coaches. After much adjustment it was decided to CON some units to a more time-tested GM-built 6-cycle diesel engine. During this process, coaches were REN to define their identity, thus 21 of the coaches became 2800s. OOS 1965–66. Sold for SCR as follows:

2700, 2716, 2720, 2738 • Kubentz 1965
2701    CON GM engine, REN 2801 8-24-55, Ours 1968, preserved at Ohio Museum of Transportation, Orrville
2702, 2703, 2705, 2708, 2709, 2714, 2726, 2727, 2733, 2735 • Ours 1965
2704    CON GM engine, REN 2804 5-2-56, Kricfalusy 1967
2706    CON GM engine, REN 2806 6-27-57, Ours 1968
2707    operated in "Parade of Progress" 4-27-52, CON GM engine, REN 2807 8-30-58, Rearich 1965
2710    CON GM engine, REN 2810 4-6-56, Ours 1968
2711    CON GM engine, REN 2811 7-27-55, Ours 1968
2713, 2725, 2746 • Kricfalusy 1967
2715    CON GM engine, REN 2815 7-24-58, Rearich 1965
2717    CON GM engine, REN 2817 3-1-63, City of Cleveland Safety Division 1968
2718    CON GM engine, REN 2818 2-22-56, Ours 1968
2719    CON GM engine, REN 2819 4-20-55, Ours 1968
2721    CON GM engine, REN 2821 8-9-56, Ours 1968

Bus 2729, Mack, 1951. *(Cleveland Transit photo, Blaine Hays Collection)*

2722, 2747 • Rearich 1965
2723    CON GM engine, REN 2823 2-4-59, Ours 1968
2724    CON GM engine, REN 2824 10-2-58, Kricfalusy 1967
2728, 2729, 2749 • R & S Metal 1967
2730    CON GM engine, REN 2830 6-18-56, Ours 1968
2731    CON GM engine, REN 2831 2-12-58, CON Power & Plant workshop and REN P-150, Dyer 1971
2732    CON GM engine, REN 2832 3-1-63, sold City of Cleveland Safety Division 1968
2734    CON GM engine, REN 2834 6-20-56, Ours 1968
2736    CON GM engine, REN 2836 10-27-58, Ours 1968
2737    CON GM engine, REN 2837 1-10-63, Ours 1968
2739    CON GM engine, REN 2939 3-30-59, Ours 1968
2741    SCR for parts
2742    CON GM engine, REN 2842 8-20-57, Ours 1968
2743    CON GM engine, REN 2843 4-19-56, Ours 1968
2744    CON GM engine, REN 2844 4-11-58, Ours 1968
2745    CON GM engine, REN 2845 9-9-55, CON Power and Plant workshop, REN P-149, Ours 1968
2748    CON to storage bin at Superior Station, sold Kubentz 1966

Bus 2983, White, 1952. *(Cleveland Transit photo, Blaine Hays Collection)*

**Fleet Statistics: White, 2947–2999, Fleet 29**

BLDR: White; LEN: 39' 4$^{13}/_{16}$"; seats: 50; model: DSL1150DW; W: 20,890 lbs; engine: 6 CYL diesel; capacity fuel tank: 85 gals; front and center doors; 2950–2999 D 4-1952 to 5-1952. ACQ to replace streetcar service on Euclid Avenue; demonstraters 2947–2949 ACQ 1953 for eliminating streetcars on Madison line; OOS 1965. Sold as follows:

2947, 2948, 2950, 2951, 2953 2955, 2956, 2957, 2967, 2968 • Maple Heights Transit 1967
2949, 2989 • Kubentz 1967
2950 operated in "Parade of Progress" 4-27-52
2952, 2969, 2979 • Perkins 1967
2954, 2963, 2971, 2973, 2987 • Ours 1967
2958, 2976 • Jones & Laughlin Steel 1967
2959 Whitman 1967
2960, 2962, 2966, 2974, 2975, 2980, 2996 • Kricfalusy 1967
2961, 2977, 2988, 2998 • SCR by CTS for parts
2994 R & S Metal 1967
2965, 2970, 2972, 2978, 2981, 2982, 2984, 2990, 2991, 2993, 2995, 2997 • Rearich 1965
2983, 2986 • Essie Radio Corp. 1967
2985 ACC 9-17-56, SCR 10-23-56

2987    sold Ours 1967, preserved at Ohio Museum of Transportation, Orrville

2999    CON tire bus and REN M-95 1964, sold Wheatly 1968

Note: Known as "teck-neck" buses because of an unusual knocking sound in engine that sounded like "teck-neck, teck-neck."

### Fleet Statistics: GM, 400–449, Fleet 04

BLDR: GM; LEN: 39' 9"; seats: 52; model: TDH 5103; W: 19,200 lbs; engine: 6 CYL diesel; capacity fuel tank: 85 gals; front and center doors; D 2-53; ACQ to replace streetcars on E. 55th St; all OOS 1971. Sold for SCR as follows:

400, 401, 403, 405, 426, 427, 432, 436 • J. Ours 1970–71
402, 418 • Seniors of Ohio 1971

Units from the GM "old look" bus fleets provided riders with leather seating and more aisle space.
*(Cleveland Transit photo, Blaine Hays Collection)*

404, 435 • Dyer
406, 413, 416 • Maple Heights Transit 1970
407   Tabor 1971
408   Calver Temple 7-19-73
409, 445, 447 • Snider 1972
410, 440 • Cleveland Board of Education 1972
411, 412, 414, 415, 423, 429 • Del-Val Coach 1972
417, 420, 430 • Rudy's Auto Wrecking 1971 (430 preserved at Ohio Museum of Transportation, Orrville)
419, 421, 422 • Mravec 1971
420   OOS 2-1-71, sold Rudy's Auto Wrecking 9-2-71
424   Emery R. Brosek 3-2-72
425, 431, 442, 443, 444 • sold City of Ashtabula 1971
428, 437 • Park Synagogue 1972
433   Albers and Fitzsimmons 1972
434, 438, 439 • Tri-County Assembly of God 1973
441   Ontario Stone Corp. 1972
448   Farmer 1972
449   Erie Amvets

### *Fleet Statistics: GM, 3100–3104, Fleet 10*

BLDR: GM; LEN: 30'; seats: 37; model: TDH-3714; engine: 4 CYL diesel; capacity fuel tank: 85 gals; front and center doors; D 9-23-53; sold J&S Trailways 12-1967.

### *Fleet Statistics: GM, 300–353, Fleet 03*

BLDR: GM; LEN: 35' 2"; seats: 45; model: TDH 4512; engine: 6 CYL diesel; front and center doors, center doors passenger-operated push-out type; D between 1-1956 and 4-1956. Some sold to Hausman Bus Sales for resale to Akron Transportation on 3-6-69: 300–302, 305, 308, 313, 316, 325, 333, 335, 336, 339, 341, 344.

311   ACC fire, SCR 2-25-60
349   41 very high-backed seats
350   D 10-1953, FN 3425, REN 350 3-7-57, GM-TDH 4512, no rear door (demonstrator ACQ for eliminating streetcars on Madison line), painted for and permanently assigned to Halle Brothers downtown department store, was replaced later by 678, sold Hausman Bus Sales 3-6-69
351   BLT 1951 for Indianapolis as number 1027, D 6-1961 from Broadview Bus Lines, FN 9, GM-TDH 4509, sold St. John's Transit Co. 1-1965

Bus 327, GMC, 1956, with push-out center doors. *(Cleveland Transit photo, Blaine Hays Collection)*

352    BLT 1953, D 6-1961 from Broadview Bus Lines, FN 10, GMTDH 4509, sold to Cleveland Southeastern 1-1965 and REN 4062

353    BLT 1954, D 6-1961 from Broadview Bus Lines, FN 11, only Broadview bus with center door, GM-TDH 4509, sold to Hausman Bus Sales 3-6-69

### Fleet Statistics: GM, 450–525, Fleet 9

BLDR: GM; LEN: 39' 8"; seats: 51; model: TDH-5105; W: 19,200 lbs; engine: 6 CYL diesel; front and center doors; tire-rack-equipped buses: 450–455, 467–467, 500–503; D 1-1954 and 2-1954; ACQ to replace streetcars on Madison line; all OOS 1970–71. Coaches sold as follows:

450   Hines 1971
451, 453, 455, 458, 462, 463, 464, 466, 467, 470, 476, 477, 479, 482, 486, 492 • Dayton-Western Motors 1972
452, 469, 485, 494, 515 • Dyer 7-22-71
454, 457, 472, 483, 500 • Ours 1970 (500 CON to Zoo Bus in 1950s)
456, 524 • Maple Heights Transit 1970
459, 468 • Smith 1972
460, 471, 499 • Charles Bus Lines 1971
461   Rudy's Auto Wrecking 1971
473, 478, 480 • Snider 1972
474   Abdul N. Ali 1971
475, 488 • S & M Santee 1971
481, 484, 487, 493, 497, 501, 503–506, 517–518, 520, 523 • Tri-County Assembly of God 1972
490   Gerson Borges 1972
491, 511 • Mt. Pleasant Baptist Church 1972
495, 496 • Ward 1971
498   Cleveland Board of Education 1972
502, 513, 514 • Ravenna Assembly of God 1972
507   United Wholesale of Kentucky 1973
508   Sar Ru Inc 1973
509   Tabor and Fischer 1973
510   Rucker 1973
512   Central Assembly of God 1972
516   Anthony E. Stempinski 1973
519   Chapel Hill Assembly of God 1972
521, 522 • Clarence Bear 1972
525   FN 3631, sold Maple Heights Transit 6-24-71

### Fleet Statistics: GM, 700–739, Fleet 19

BLDR: GM; LEN: 30'; seats: 37; model: TDH-3714; W: 16,850 lbs; engine: 4 CYL diesel; capacity fuel tank: 85 gals; front and center doors; D 8-1955. All sold to Conestoga Transportation Co. 3-6-69 except the following:

704, 728 • City Transit Lines 1966
705–712, 722, 723 • United Transit Lines, Puerto Rico, 1966
707   CON Christmas bus 1950s, United Transit Lines Puerto Rico 2-10-66
712   CON Christmas bus 1950s, United Transit Lines Puerto Rico 2-10-66
715, 732, 733 • Springfield City Lines 1966
734   PTD for Boy Scouts in 1960s
719   CON to Fun Tour bus 1950s

### Fleet Statistics: GM, 530–570, Fleet 5

BLDR: GM; LEN: 39' 8"; seats: 51; model: TDH-5105; W: 19,200 lbs; engine: 6 CYL diesel; 530–533, 560–563 tire-rack equipped; 569, 570 (FN 12) ACQ from Broadview Bus Lines 6-1961; front and center doors; D 10-1955 to 12-1955; all OOS 1971. Sold as follows:

530, 534, 540, 547, 557 • Catawese Coach Lines 1973
531, 539 • Ward 1971
532, 536, 553 • Rudy's Auto Wrecking 1971
533, 544 • Charles Bus Lines 1973
535, 537, 538, 546, 551 • Snider 1973
541, 545, 550, 558, 563, 568 • Dyer 1971
542    Tri-County Assembly of God 1972
543    Castellano 1971
548    Tabor 1973
549    Central Assembly of God of Mentor 1972
552, 561 • Ours 1971
554, 556, 560, 567 • Western Reserve Transit 1975, 556 CON to Zoo bus in 1950s
559    Maple Heights Transit 1971
562    Dayton Western Motors 1971
564    City of Cleveland Office of Consumer Affairs 1973
565, 566 • City of Ashtabula 1973
569    Cleveland Board of Education 1972

### Fleet Statistics: GM, 1700–1739, Fleet 17

BLDR: GM; LEN: 39' 8"; seats: 51; model: TDH-5105; W: 19,200 lbs; engine: 6 CYL diesel; front and center doors, center doors were passenger-operated push-out type; D between 7-10-57 and 7-30-57; 1700, 1708–1711 tire-rack-equipped buses; all OOS 2-1-71. All sold to Atlanta Transit System between 6-15-72 and 9-7-72 except the following:

1700    Maple Height Transit 1971
1705, 1707, 1712, 1715, 1718, 1720, 1724, 1729, 1730, 1735, 1739 • Western Reserve Transit 1975
1713    Rudy's Auto Wrecking 1971
1722    B. Dyer 1971
1736    Lake Front Lines 1970

### Fleet Statistics: GM, 1800–1830, Fleet 18

BLDR: GM; seats: 51; model: TDH-5105; W: 20,000 lbs; engine: 6 CYL diesel; front and center doors, center door was passenger-operated push-out type; tire-rack equipped: 1825–1829; 1830 (FN 15) ACQ from Broadview Bus Lines 6-1961; all other buses D 11-1958. Sold as follows:

1800, 1801, 1810–1812, 1814, 1819, 1823 • Western Reserve Transit 1975
1802    Dyer Wrecking 1971
1803    Leader Metal & Supply 1974
1805, 1807, 1808, 1813, 1824 • Williamstown Irrigation 1978
1809    Tri-County Assembly of God 1972
1818    PTD for Cleveland State University 1971, Tri-County Assembly of God 1972
1827    Tri-County Assembly of God 1972
1816, 1817, 1820, 1826 (given to Maple Heights Transit in settlement of accident claim 6-12-74), 1830 • Maple Heights Transit 1971
1821    Beaver Valley Motor Coach 1961
1828    leased to City of Ashtabula in 1974, sold 1977

### Fleet Statistics: GM, 600–699, Fleet 06

BLDR: GM; LEN: 40'; seats: 53; model: TDH-5301; W: 20,555 lbs; engine: 6 CYL diesel; capacity fuel tank: 95 gals; front and center doors; D 9-1959 to 11-1959. Cleveland's first order of "New Look" buses, also known as "Modernaire" buses: coaches 695–697 were spare-tired, over-the-road Deluxe; 698–699 were AC but not particularly successful because 6-cylinder engine was not sufficiently powerful. Some were paraded through downtown on 10-2-59 to show them off to the public. They were the first dark-blue buses, PTD same as rapid transit cars. Coach 678 PTD light-green and was permanently assigned to the Halle Brothers department store downtown. It was later replaced by bus 1507. Some of the 600s lasted into the RTA era as follows: 600–616, 618–638, 640–644, 646–648, 650–653, 655–661, 663–669, 671, 673–698. As a result of its operating agreements with the suburban bus companies, RTA leased the following buses:

602, 606, 607, 608 • Maple Heights Transit
637, 638 • Garfield Heights
619, 623, 624, 626, 628, 629, 630, 631, 632, 633, 634 • North Olmsted
627    Brecksville
641    preserved at the Ohio Museum of Transportation, Orrville
660    training coach at Brooklyn
669    training coach at Hayden
685    training coach at Triskett
694    training coach at Woodhill

Bus 678, Fishbowl, GMC, 1959, assigned to the Halle Brothers Company, provided well-known free service to the Playhouse Square store. *(Cleveland Transit photo, Blaine Hays Collection)*

### *Fleet Statistics: GM, 800–832, Fleet 08*

BLDR: GM; LEN: 40'; seats: 53; model: TDH-5301; W: 19,766 lbs; engine: 6 CYL diesel; capacity fuel tank: 95 gals; front and center doors; ACQ 1960; D between 10-12-60 and 11-1-60; all transferred to RTA except 800, 805, 831.

- 800  stripped for parts 4-1972
- 805  sold Ours 5-7-71
- 830  ACQ from Broadview Bus Lines in 6-1961, FN 16, had no rear door
- 831  ACQ from Broadview Bus Lines in 6-1961, FN 17, had no rear door, sold to Maple Heights Transit 3-10-71
- 832  ACQ from Berea Bus Lines in 2-1968, no rear door, FN 128, REN 4-17-68, sold to Pielet Bros. 10-26-76

### *Fleet Statistics: Flxible, 1400–1429, Fleet 14*

BLDR: Flxible-Twin Coach; LEN: 40'; seats: 53; model: FlxF2D6V-401-1; W: 21,600 lbs; engine: 6 CYL diesel; capacity fuel tank: 95 gals; front and center doors; D 9-1961; all transferred to RTA. These were of the "New Look" type but had a squared-off slanted window unlike any other; they were some of the very last buses built by the combined Flxible-Twin Coach and had the combined builders' plate. Buses 1427–1429 were Deluxe with tire racks for over-the-road service.

### *Fleet Statistics: Flxible, 1500–1529, Fleet 15*

BLDR: Flxible; LEN: 40'; seats: 53; model: Flx-F2D6V-401-1; W: 20,930 lbs; engine: 6 CYL diesel; capacity fuel tank: 95 gals; front and center doors; D 7-1962 and 8-1962; all transferred to RTA. 1507 PTD dark-green and was permanently assigned to Halle Brothers department store downtown and operated a daily shuttle service between Public Square and Halle's. 1515 was converted into a "Mother Bus" in 1982 to serve as a repository for repair parts for on-the-road emergency repairs. 1523 was destroyed by fire during labor unrest, a dispute with management 2-1979.

### *Fleet Statistics: Flxible, 1600–1664, Fleet 16*

BLDR: Flxible; LEN: 40'; seats: 53; model: FlxF2D6V-401-1; W: 20,950 lbs; engine: 6 CYL diesel; capacity fuel tank: 95 gals; front and center doors; D 9-1963; all transferred to RTA except 1656, which was sold to Ours 8-30-71 for scrap after an accident.

Bus 1664, Flxible, 1963. *(Cleveland Transit photo, Blaine Hays Collection)*

### *Fleet Statistics: GM, 1900–1959, Fleet 19*

BLDR: GM; LEN: 40'; seats: 53; model: TDH-5303; engine: 6 CYL diesel; capacity fuel tank: 95 gals; front and center doors; D 4-1963; all transferred to RTA. This fleet replaced last of the trackless trolleys on east side lines 6-15-63. In 1972 coaches 1910–1939 PTD orange with white and black stripes for downtown loop service and dubbed "orange aid for downtown."

### *Fleet Statistics: GM, 1000–1024, Fleet 10*

BLDR: GM; LEN: 40'; seats: 53; model: TDH-5303; engine: 6 CYL diesel; front and center doors; D 8-1964; all transferred to RTA.

### *Fleet Statistics: Flxible, 1300–1324, Fleet 13*

BLDR: Flxible; LEN: 40'; seats: 52; model: FlxF2D6V-401-1; engine: 6 CYL diesel; capacity fuel tank: 95 gals; front and center doors; D 8-1964 and 9-1964; all transferred to RTA.

### *Fleet Statistics: GM, 1100–1144, Fleet 11*

BLDR: GM; LEN: 40'; seats: 53; model: TDH-5303; engine: 6 CYL diesel; front and center doors; D 8-1965 and 9-1965; all transferred to RTA.

### *Fleet Statistics: Flxible, 1200–1209, Fleet 12*

BLDR: Flxible; LEN: 30'; seats: 37; model: Flx FD47-331-1; engine: 4 CYL diesel; capacity fuel tank: 95 gals; front and center doors; D 10-13-65 and 10-14-65, for loop service (PTD bright-red with a large "L"

Bus 1202, Flxible Flxette, 1965, purchased for loop service. *(Cleveland Transit photo, Blaine Hays Collection)*

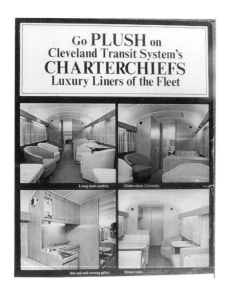

on front to stand out from other buses and to proclaim "Loop"). Operation was largely unsuccessful because operators were paid the same wages but had less passenger capacity. Loop buses were good hauling routes, and passengers found themselves jammed into these little Flxette coaches. All in this fleet were later PTD blue and operated on light hauling routes out of Brooklyn Station. All coaches transferred to RTA and sold to Williamstown Irrigation on 4-11-78. 1200 was leased to the University of Virginia in 1974 and sold to Williamstown with the others.

### Fleet Statistics: GM, 2100–2159, Fleet 21

BLDR: GM; LEN: 40'; seats: 53; model: TDH-5303; engine: 6 CYL diesel; capacity fuel tank: 95 gals; front and center doors; D 8-1967 except number 2100, which came in 7-1967; all transferred to RTA. They were distinctive in that they had a sloping white-painted band on the roof that curved up just before the front door. 2105 preserved at the Ohio Museum of Transportation, Orrville

### Fleet Statistics: Charterchief, 2400–2402, Fleet 24

2400 is a long Twin 52-SD-2, originally a propane coach. See 3785 for details. 2401 and 2402 came 3-10-69 to replace the Twin. They are GMC-T8H-5305A models with AC and fully decked-out interiors including refrigerator, stove, tables, plush chairs. None of the Charterchiefs had rear doors. 2401 and 2402 were transferred to RTA.

### Fleet Statistics: Flxible, 3000–3099, Fleet 30

BLDR: Flxible; LEN: 40'; seats: 53; model: Flx-111-CC-D5; engine: 6 CYL diesel; front and center doors; D 1-1969 to 3-1969; all transferred to RTA. These were a stripped-down model of "basic" bus, no frills and hard plastic seats. 3028 preserved at Ohio Museum of Transportation, Orrville.

### Fleet Statistics: Flxible, 100–199, Fleet 71

BLDR: Flxible; LEN: 40'; seats: 50; model: Flx-111-CC-D61; W: 11 tons; engine: 8 CYL diesel; front and center doors; 25 equipped with spare tire and bucket "Clevelander" seats for long charters; D 1971. This fleet inaugurated two firsts for CTS. It was the first large fleet of AC buses, an integral part of a new marketing thrust being implemented by General Manager Robert T. Pollock. It was also the first

Bus 181, Flxible, 1971, with water bumpers. *(Cleveland Transit photo, Blaine Hays Collection)*

complete change in color for the system since the introduction of the blue schemes in the mid-1950s. The all-white with blue and green belt rail stripes began the trend of white as the base color for CTS and later RTA vehicles. All transferred to RTA.

### *Fleet Statistics: GM, 2500–2651, Fleet 25*

BLDR: GM; LEN: 40'; seats: 50; model: T8H-5307-A; engine: 8 CYL diesel; front and center doors; D 1-1975 and 2-1975; all transferred to RTA. These had AC and plush bucket seats, perfect for ushering in the new RTA era. As a result of its operating agreements with the suburban bus companies, RTA leased the following in 1975:

2587, 2588 • Maple Heights
2584, 2585, 2586 • Garfield Heights
2649, 2650, 2651 • North Olmsted
2510    involved in serious front-end ACC
2589    Brecksville
2609    preserved at the Ohio Museum of Transportation, Orrville

Note: Bodies of 2644, 2645, and 2650 retained as parts supply for 2401–2402 Charterchief buses.

### SECTION 3
### *Buses Purchased by CTS from Other Transit Agencies*

#### *Fleet Statistics: Twin, 3786–3799, Fleet 24*

BLDR: Twin; seats: 52; 3786–3797, FN 403–414, BLT 1947; 3798–99, FN 501–502, BLT 1950, were gasoline-fueled and ACQ by Lakewood Rapid Transit in 1947 and ACQ by CTS upon takeover of that system in 1954. Also ACQ by CTS from Lakewood were Twin (1937) 31R buses 201–205; Twin (1937) 35R 301–302; Twin (1941) 35G 303; and White (1946) 788 401–402; these were not used in CTS service and were sold.

| | |
|---|---|
| 3786 | St. Stephen's, Niles, 1963 |
| 3787 | M. Kricfalusy 1964, preserved at Ohio Museum of Transportation, Orrville |
| 3790 | sold Kubentz Iron Co. 1960 |
| 3791 | M. Kricfalusy 1964 |
| 3792 | sold without board authorization as substitute for 3681 in 1963 |
| 3793 | St. Helen's Church, Newbury |
| 3795 | OOS 2-1964, Cleveland Southeastern Trails 1964 |
| 3796, 3797 · Jones & Laughlin | |
| 3798 | CON to Safety Department bus 1960 |

#### *Fleet Statistics: Berea Bus Lines Coaches*

All coaches D to CTS 1-1-68. "Ours" is a scrapyard called Jacob Ours Truck Parts, New Brighton, PA; (S) stands for school bus.

| | |
|---|---|
| 30 | BLT 1940, White 788-1 suburban, sold Trolleyville USA 5-17-68 |
| 42R | BLT 1941, White 798-1 transit, sold Cambridge Motor Coach 6-20-68 |
| 48 | BLT 1945, White 798-10 suburban, sold Trolleyville USA 5-17-68 |
| 50 | BLT 1945, White 798-10 suburban, sold Hopkins Airport Limo Service 5-17-68 |
| 52 | BLT 1946, White 798-10 suburban, sold Trolleyville USA 5-17-68 |
| 54 | BLT 1946, White 798-10 suburban, sold Trolleyville USA |

| | 5-17-68, sold to Hale Farm, given to NORM 1986, sold to Ohio Museum of Transportation 1993 |
|---|---|
| 58, 64, 70, 80 • | BLT 1947, White 798-10 suburban, sold 3900 Corp. 4-1968 |
| 66 | BLT 1948, White 798-10 suburban, sold Ambridge Motor Coach 6-20-68 |
| 78 | BLT 1951, White 1144-S suburban, sold Trolleyville, USA, 6-20-68, sold to Hale Farm, given to NORM 1986, sold to Ohio Museum of Transportation 1993 |
| 88 | BLT 1954, White (S) WC16B Superior, sold Ours 7-11-68 |
| 90 | BLT 1954, White (S) WC16B Wayne, sold Ours 7-11-68 |
| 102 | BLT 1956, White (S) WC16B Wayne, sold Ours 7-11-68 |
| 104 | BLT 1956, White (S) EC16B Wayne, sold John F. Kennedy High 7-11-68 |
| 106 | BLT 1950, White (S) T-1140-1 White, sold Glady 7-11-68 |
| 108, 110, 112, 114 • | BLT 1950, White (S) T-1140-1 White, sold Wheatly 7-11-68 |
| 116, 118 | BLT 1957, GM (S) T3021 Wayne, sold Berea Board of Education 6-1-68 |
| 120, 122 | BLT 1957, GM (S) TA3021 Wayne, sold same |
| 128 | BLT 1959, GM TDH-5301 transit, REN CTS 832 (fleet 8) on 4-17-68 |
| 130, 132, 134, 136 • | BLT 1959, White (S) WC2000B Wayne, sold Berea Board of Education |
| 150, 152, 154, 156 • | BLT 1960, Ford (S) B-75NQU Wayne, sold same |
| 166 | BLT 1950, White 798-1 transit, sold Weitzel 6-20-68 |
| 168 | BLT 1963, GM PD-4106 (AC) intercity, sold McGuigan 3-6-68 |
| 170 | BLT 1963, Flxible DDI-63-29 (AC) intercity, sold Yellow Cab Co. 3-13-68 |
| 172 | BLT 1964, GM PD-4106 (AC) intercity, sold Cleveland-Lorain Hwy Coach 3-7-68 |
| 174, 176 | BLT 1964, GM SMD-5302 suburban, sold Cleveland Southeastern 6-20-68 |
| 178, 180 | BLT 1965, GM (S) TDSPA501-9 Wayne, sold Berea Board of Education 6-1-68 |
| 182 | BLT 1965, GM SMD-5302 (AC) suburban, sold Cleveland Southeastern 4-1968 |
| 184 | BLT 1966, GM (S) DSA4022 Wayne, sold Berea Board of Education 6-1-68 |
| 188 | BLT 1966, GM SDH-5302 suburban, REN CTS 2305 (fleet 23) 4-17-68 |
| 190 | BLT 1966, GM SDH-5302 suburban, REN CTS 2306 (fleet 23) 4-17-68 |
| 192 | BLT 1966, GM SDM-5302 (AC) suburban, sold Blue Bird Coach Lines 4-1968 |

194, 196, 198 • BLT 1966, GM (S) DSA4022 Wayne, sold Berea Board of Education 6-1-68

### *Fleet Statistics: Mack, 2000–2007, Fleet 20*

BLDR: Mack; seats: 51; model: Mack C 49DT; engine: 6 CYL diesel; front and center doors; 2000 and 2001 D 12-12-57; ACQ to compare with 1700–1701 GM type, but Mack went out of business before test was completed. CTS purchase of the Redifer Bus Company yielded 20 buses, of which only 6 were used by CTS. 2002–2007 BLT 1956 for Redifer, FN 501–507. The others, 8 1958 Macks and 12 Twins were sold immediately. Those retained by CTS were later sold as follows:

| | |
|---|---|
| 2000, 2001 | Avenue B and East Broadway Transit System |
| 2002 | OOS 3-2-65 |
| 2003–2006 | Ours 1967 |
| 2007 | Rearich 1966 |

### *Fleet Statistics: GM, 2300–2306, Fleet 23*

BLDR: GM; LEN: 40'; seats: 49; model: SDH-5302; engine: 6 CYL diesel; capacity fuel tank: 95 gals; front and center doors; 2300–2304 were AC, D 6-22-65; sold to New York Bus Tours 3-18-71. 2305 had no center door and came as number 188 from Berea Bus Lines 2-1-68. 2306 had no center door, sported a card table in the back, and came as number 190 from Berea Bus Lines 2-1-68. They had posh high-backed seats. They operated an express service for GE employees for several years, and the GE logo appeared on the roller sign. 2305–2306 were transferred to RTA.

| | |
|---|---|
| 2306 | ACQ by Ohio Museum of Transportation but later sold due to poor condition |

### Motor Coach Service Units

Some double-deck frames (numbers unknown) became salt spreaders as follows:

| | |
|---|---|
| A-9 | salt spreader, assigned to Berea Station |
| A-10 | salt spreader, assigned to Berea Station |
| A-11 | salt spreader, assigned to Superior Station |
| A-12 | salt spreader, assigned to Brooklyn Station |
| A-19 | salt spreader, assigned to Superior Station |

Service bus, tire truck. *(Cleveland Railway photo, Blaine Hays Collection)*

| | |
|---|---|
| A-20 | salt spreader, assigned to Superior Station |
| A-21–A-25 | salt spreaders, D 2-25-47 (not made from double deckers) |
| A-21 | assigned to Superior Station |
| A-22 | assigned to Superior Station |
| A-23 | assigned to Windermere Station |
| A-24 | assigned to Lorain Station |
| A-25 | assigned to Brooklyn Station |
| ★ | tire truck |
| 6 | service truck |
| 8 | revenue truck |
| B-4 | St. Clair Station (from photo) |
| B-5 | Berea Garage |
| B-13 | gardener coach |

Note: Except for those mentioned above, the "B" series coaches were not included in inventory. It is believed that there was one assigned to each station and the remainder in the "B" series were trucks.

# SECTION 4
## *Buses of Regional Transit Authority Acquired from Other Operators and Purchased New*

The Greater Cleveland Regional Transit Authority was created on December 30, 1974, by action of the Cleveland City Council and the Cuyahoga County Board of Commissioners. A governing board of trustees for the new authority was appointed on January 29, 1975.

On May 21, 1975, a "Memorandum of Understanding" covering the transfer of CTS to RTA and outlining anticipated service improvements was signed by city, county, and CTS officials. Implementation of the memorandum was dependent upon a positive vote by the electorate to increase the county sales tax by one cent. The July 22 vote brought overwhelming approval.

On September 5 the Cleveland Transit System and the Shaker Heights Rapid Transit were permanently transferred to RTA control, and employee passes on either system were honored by the other. At midnight on October 5, the new reduced-fare structure that RTA had promised went into effect on all area transit services, including the five suburban bus systems, which by then had operating and subsidy agreements with RTA.

This section of the roster covers the buses that RTA acquired from predecessor agencies and those that it purchased new. Equipment in the roster appears in the order that it was acquired. Dates equipment was acquired from suburban systems are the same as the dates of takeover as noted below.

The Regional Transit Authority represents the amalgamation of eight formerly independent transit operations. Some were taken over directly by RTA; others started with operating agreements with RTA but were later fully absorbed into the Authority. Two (North Olmsted and Maple Heights) have remained independent but operate with an RTA service agreement. The agencies and how they were affected are as follows:

*Brecksville Road Transit,* started 1946, ACQ RTA 7-1-84
*Brothers-in-Christ (BIC) Transit,* started 1971, ACQ RTA 9-5-75
*City of Euclid (Transit Division),* started 8-16-35, ACQ RTA 7-1-79
*Cleveland Transit System,* started 4-28-42, ACQ RTA 9-5-75
*Garfield Heights Coach Lines (Transit Division),* started 4-1-32, ACQ RTA 1-21-80
*North Olmsted Municipal Bus Lines,* started 1931, service agreement 10-5-75
*Maple Heights Transit System,* started 1935, service agreement 10-5-75

Bus 8114, AM General, 1978. *(AM General photo, Blaine Hays Collection)*

*City of Shaker Heights Department of Transportation,* started operations 12-17-13, ACQ RTA: 9-5-75

RTA has provided buses to the two independent operators (Maple Heights and North Olmsted) as follows:

*Maple Heights Transit:*
10-31-75: all CTS: 602, REN 5001, 606, REN 5002, 607, REN 5003, 608, REN 5004, 2587, 2588
6-30-96: all RTA: 8376–8377, 8589–8594, 8604, 8605, 8905–8906, 9065–9069, 9255–9258

*North Olmsted Municipal:*
10-31-75: all CTS: 623, 624, 628, 629, 631, 632, 2649, 2650, 2651
6-30-96: all RTA: 8374–8375, 8588, 8595–8603, 8901–8904, 9070–9074, 9151–9156, 9201–9202, 9305

## RTA Buses Acquired from Other Operators

| Series | Fleet | Manufacturer | Acquired from | Seats | Cylinder | Built | Number | Note | Out of Service |
|---|---|---|---|---|---|---|---|---|---|
| 1800–1829 | 18 | GM TDH-5105 | CTS | 51 | 6 | 1958 | 12 | 1 | 1975 |
| 600–699 | 06 | GM TDH-5301 | CTS | 53 | 6 | 1959 | 88 | 1 | 9-1984 |
| 832 FN 128 | 08 | GM TDH-5301 | CTS | 51 | 6 | 1959 | 1 | 1 | 6-1983 |
| 800–831 | 08 | GM TDH-5301 | CTS | 53 | 6 | 1960 | 29 | 1 | 6-1983 |
| 1400–1429 | 14 | Flxible F-2D6V-401-1 | CTS | 53 | 6 | 1961 | 30 | 1, F | 1985 |
| 1500–1529 | 15 | Flxible F-2D6V-401-1 | CTS | 53 | 6 | 1962 | 30 | 1 | 1985 |
| 1600–1664 | 16 | Flxible F-2D6V-401-1 | CTS | 53 | 6 | 1963 | 64 | 1 | 1985 |
| 1900–1959 | 19 | GM TDH-5303 | CTS | 53 | 6 | 1963 | 60 | 1 | 11-1984 |
| 1000–1024 | 10 | GM TDH-5303 | CTS | 53 | 6 | 1964 | 25 | 1 | 1985 |
| 1300–1324 | 13 | Flxible F-2D6V-401-1 | CTS | 52 | 6 | 1964 | 25 | 1 | 1985 |
| 1100–1144 | 11 | GM TDH-5303 | CTS | 53 | 6 | 1965 | 45 | 1, H, I | 3-1989 |
| 1200–1209 | 12 | Flxible FD47-331-1 | CTS | 37 | 4 | 1965 | 10 | 1 | 1978 |
| 2305 FN 188 | 23 | GM TDH-5302 | CTS | 53 | 6 | 1966 | 1 | 1 | 11-1984 |
| 2306 FN 190 | 23 | GM TDH-5302 | CTS | 49 | 6 | 1966 | 1 | 1 | 1985 |
| 2100–2159 | 21 | GM TDH-5303 | CTS | 53 | 6 | 1967 | 60 | 1, D | 9-1989 |
| 2401–2402 | 24 | GM TDH-5305A | CTS | 28 | 8 | 1969 | 2 | 1 | |
| 3000–3099 | 30 | Flxible 111-CC-D5 | CTS | 53 | 6 | 1969 | 100 | 1 | 6-1988 |
| 100–199 | 71 | Flxible 111-CC-D61 | CTS | 50 | 8 | 1971 | 100 | 1 | 6-1988 |
| 2500–2651 | 25 | GM T8H-5307A | CTS | 50 | 8 | 1975 | 152 | 1, L | 9-1992 |
| 1–3 | | | B.I.C. | | | | 3 | 2 | 1&2 1977; 3 5-1979 |
| 9000–9001 | 90 | Dodge CRT | city-CTS | 15 | 8 | 1975 | 2 | 3, 12 | |
| 9100–9111 | 91 | Argosy (Chevy) CRT | city-CTS | 15 | 8 | 1975 | 12 | 3, 12 | |

CRT NET coach, Argosy, 1973.
*(Blaine Hays photo)*

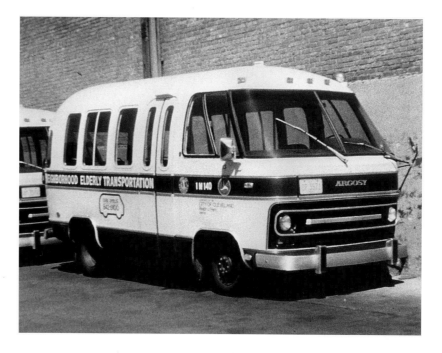

| Series | Fleet | Manufacturer | Acquired from | Seats | Cylinder | Built | Number | Note | Out of Service |
|---|---|---|---|---|---|---|---|---|---|
| 9200–9204 | 92 | Dodge CRT (county) | FN RR1–RR5 | 8 | | 1975 | 5 | 3 | 1992 |
| 4 FN 134 | | GM TDM-4509 (suburban) | CLC | 45 | | 1953 | 1 | 13 | 10-1977 |
| 5 FN 141 | | GM SDM-4501 (suburban) | CLC | 45 | | 1961 | 1 | 13 | 10-1977 |
| 2 | | | CLC | | | | 1 | 13 | 5-1979 |
| 1335–1339 | | GM TDH-4509 | Euclid | 45 | 6 | 1952 | 2 | 7, B | 3-1981 |
| 1004 | | GM TDH-4509 | Euclid | 45 | 6 | 1953 | 1 | B | 6-1982 |
| 1440–1445 | | GM TDH-4512 | Euclid | 45 | 6 | 1953 | 4 | 7, C | 1983 |
| 1505–1510 | | GM TDH-4512 | Euclid | 45 | 6 | 1956 | 6 | 7, C | 1983 |
| 1746–1754 | | GM TDH-5105 | Euclid | 51 | 6 | 1957 | 9 | 7, C | 1983 |
| 2114–2115 | | GM TDH-4517 | Euclid | 45 | 6 | 1961 | 2 | 7 | 1984 |
| 2330–2333 | | GM TDH-5304 | Euclid | 53 | 6 | 1963 | 4 | 7 | 1983 |
| 2626–2629 | 27 | GM TDH-5304 | Euclid | 53 | 6 | 1966 | 4 | 7 | 1985 |
| 3055–3067 | | GM T6H-5306 | Euclid | 53 | 6 | 1970 | 13 | 7 | 1986 |
| 700–701 | 07 | GM TDH-4517 | Euclid | 45 | 6 | 1961 | 2 | 7 | 6-1982 |
| 5535–5543 | | GM TDH-5301 | Garfield | 55 | 6 | 1959 | 2 | 9 | 6-1982 |
| 5595, 5598, 5600 | | GM TDH-5301 | Garfield | 55 | | 1959 | 3 | 9 | 6-1982 |
| 5545–5547 | | GM TDH-5301 | Garfield | 55 | 6 | 1960 | 2 | 9 | 6-1982 |
| 5550 | | GM TDH-5301 | Garfield | 55 | 6 | 1962 | 1 | 9 | 6-1982 |
| 5555–5556 | | GM TDH-5303 | Garfield | 55 | 6 | 1963 | 2 | 9, F | 9-1982 |
| 5558–5560 | | GM TDH-5303 | Garfield | 55 | 6 | 1964 | 2 | 9 | 1983 |
| 5564–5565 | | GM TDH-5303 | Garfield | 55 | 6 | 1965 | 2 | 9 | 1983 |
| 5566, 5568 | | GM TDH-5303 | Garfield | 55 | 6 | 1966 | 2 | 9, F | 1983 |
| 5569–5570 | | GM TDH-5303 | Garfield | 55 | 6 | 1966 | 2 | 9 | 1983 |
| 839–899 | 18 | GM TDH-5303 | Marta | 53 | 6 | 1963 | 50 | 4, 10, E | 9-1992 |
| 6319 | 63 | GM TDH-5303 | Brecksville | 55 | 6 | 1963 | 1 | 8 | 1987 |
| 6420–6429 | 64 | GM TDH-5303 | Brecksville | | 6 | 1964 | 4 | 8 | 11-1987 |
| 6521 | 65 | GM TDH-5303 | Brecksville | 55 | 6 | 1965 | 1 | 8 | 10-1990 |
| 6530 | 65 | GM TDH-5303 | Columbus | 53 | 6 | 1965 | 1 | 14 | 10-1990 |
| 6622 | 66 | GM TDH-5303 | Brecksville | 55 | 6 | 1966 | 1 | 8 | 10-1990 |
| 6649 | 66 | GM TDH-5303 | Columbus | 53 | 6 | 1966 | 1 | 14 | 10-1990 |
| 6723 | 67 | GM TDH-5303 | Brecksville | 55 | 6 | 1967 | 1 | 8 | 1987 |
| 6804–6818 | 68 | GM T6H-5305A | Columbus | 53 | 6 | 1968 | 3 | 14 | 10-1990 |
| 6924 | 69 | GM TDH-5303 | Brecksville | 55 | 6 | 1969 | 1 | 8 | 1987 |
| 6920–6930 | 69 | GM T6H-5305A | Columbus | 53 | 6 | 1969 | 4 | 14 | 10-1990 |
| 7025–7026 | 70 | GM TDH-5303 | Brecksville | 53 | 6 | 1971 | 2 | 8 | 1987 |
| 7032–7045 | 70 | GM T6H-5305A | Columbus | 53 | 6 | 1969 | 5 | 14 | 10-1990 |
| 7551–7558 | 75 | GM T6H-5307A | Columbus | 50 | 6 | 1975 | 2 | 14 | 9-1992 |

*Buses Purchased New by RTA*

| Series | Fleet | Manufacturer | Acquired for | Seats | Cylinder | Built | Number | Note | Out of Service |
|---|---|---|---|---|---|---|---|---|---|
| 9300–9314 | 93 | Argosy (Chevy) | CRT | 12 | 8 | 1977 | 15 | 3 | 1990 |
| 9400–9414 | 94 | Argosy (Chevy) | CRT | 15 | 8 | 1977 | 15 | 3 | 1990 |
| 8000–8142 | 80 | A.M. General | | 48 | 8 | 1978 | 143 | 10, K | 10-1990 |
| 4100–4256 | 41 | GM T8H-203 | | 45 | 8 | 1979 | 157 | G, K | |
| 9500–9514 | 95 | Arc Commuter (GM) | CRT | 19 | 8 | 1980 | 15 | 3 | 1990 |
| 8201–8277 | 82 | GM T80-204 | | 47 | 6 | 1982 | 77 | 8 | |
| 9600–9614 | 56 | Transmaster (Skillcraft) | CRT | 20 | 8 | 1983 | 15 | 3 | |
| 8301–8377 | 83 | Flx-Metro | | 48 | 6 | 1984 | 77 | 17, K | |
| 9615–9629 | 56 | Transmaster (Skillcraft) | CRT | 20 | 8 | 1984 | 15 | 3 | |
| 8490–8499 | 84 | Transmaster (Skillcraft) | CRT | 20 | | 1984 | 10 | 15 | 1992 |
| 8501–8605 | 85 | Flx-Metro | | 47 | 6 | 1985 | 105 | | |
| 5730–5742 | 57 | Orion II (BIA) | CRT | 20 | 8 | 1986 | 13 | 3 | |
| 8801–8877 | 88 | Flx-Metro (Cummins engine) | | 44 | 6 | 1988 | 77 | | |
| 8901–8977 | 89 | TMC T802 | | 47 | 6 | 1989 | 77 | 20 | |
| CNG-1 | | CNG Flx-Metro (loaned 2-1990) | | 44 | 6(G) | 1990 | 1 | 5, 17 | |
| 9001–9074 | 90 | Flx-Metro (Cummins engine) | | 44 | 6 | 1990 | 74 | | |
| 9101–9176 | 91 | Flx-Metro (Cummins engine) | | 45 | 6 | 1990 | 76 | L | |
| 4701–4715 | 47 | Flx-Metro-30102 (30') | | 27 | 6(G) | 1991 | 15 | 6, 17 | |
| 4716–4720 | 47 | Flx-Metro-30102 (35') | | 27 | 6(G) | 1991 | 5 | 6, 17 | |
| 9201–9258 | 92 | Flx-Metro | | 45 | 6 | 1991 | 58 | L | |
| 5801–5820 | 58 | Champion | CRT | | 8 | 1991 | 20 | 3 | |
| 5901–5918 | 59 | Orion II (BIA) | CRT | 18 | | 1991 | 18 | 3 | |
| 9301–9305 | 93 | Flx-Suburban (Cummins engine) | | 45 | 6 | 1994 | 5 | 16 | |
| 9401–9465 | 94 | Flx-Metro | | 44 | 4(G) | 1994 | 65 | 18 | |
| 9501–9515 | 94 | Flx-Suburban | | 44 | 4(G) | 1995 | 15 | 16, 18 | |

(G) = natural-gas fueled

### RTA ROSTER NOTES

1. A total of 835 buses, from Fleet 18 (1800–1829) to Fleet 25 (2500–2651) were turned over by CTS to RTA on 9-5-75. Twelve coaches in the 1800–1829 series went to RTA but never ran, and most were sold for scrap in 1978, and all were gone by June 1979. They were: 1804–1808, 1813, 1815, 1822, 1824–1825, 1828–1829.

Bus 4131, GMC, 1979. *(GMC photo, Blaine Hays Collection)*

Bus 8969, TMC, 1989. *(TMC photo, Blaine Hays Collection)*

PART D: MOTOR COACH ROSTER · 259

Bus 9423, Flxible, 1994, at the new Harvard Garage. *(Blaine Hays Photo)*

2. The three B.I.C. buses were: 1-Flxible, 2-General Motors, 3-Continental Golden Eagle.
3. "CRT" stands for Community Responsive Transit and is the specialized service for the handicapped and Extra Lift service for the severely disabled. The original 9000–9001 and 9100–9111 fleets were purchased by the federal government during the end of the CTS era for Cleveland's Neighborhood Elderly Transportation (NET) dial-a-bus program. The 9100–9111 Argosy vans were formerly numbered 11–19 and 21–23, and the Dodge vans (9000–9001) were formerly numbered 41 and 42; REN shortly after being turned over to RTA. In later years some taxicabs were engaged in CRT service, but they were not owned by RTA. Vans in the 2500 series are leased from Lakefront Trailways and are not included in the roster.
4. Buses 839–899 were leased from Metropolitan Atlanta Rapid Transit Authority in 1981 to ease a bus shortage. They were later purchased by RTA and had the following numbers: 839–849, 851–861, 863, 865–871, 873–883, 885–888, 890–896, 898–899.
5. "CNG-1" means Consolidated Natural Gas, a standard transit 40' demonstrator bus owned by East Ohio Gas Company parent, Consolidated Natural Gas. ACQ by RTA after use in demonstration service in several cities and REN as loop bus number 4721.
6. 4701–4721 CNG coaches. Historically Cleveland's loop services were "route 47," and so prefix "47" and special paint were applied to these "loop only" buses. In 1993 coaches 4701 and 4702 were painted standard RTA colors for use on lightly traveled non-loop lines; 4703 and 4704 followed with this scheme in 1994. 4700 was experimental CNG water-going bus.

Bus CNG 1, Flxible, 1990, natural-gas bus. (*Jim Wysocki photo*)

7. City of Euclid. Upon takeover by RTA, several Euclid fleets had conflicting numbers with existing RTA fleets. Some were REN as follows: 2114–2115 became 700–701, 2626–2629 became 2726–2729, and 3055–3067 became 1200–1212. Those that weren't REN were physically identified with an "E" prefix. In fleet 1440–1445 there was no 1443 or 1444.

8. Beginning with the 8201–8277 fleet (82), RTA began the practice of having the first two digits represent the year built. This practice was extended to acquired suburban buses from Brecksville and used Columbus coaches. The second two digits of the Brecksville fleet were the original numbers as follows: (63)19, (64)20–55 seats, (64)27–53 seats, (64)28–53 seats, (64)29–53 seats, (65)21, (66)22, (67)23, (69)24, (70)25, (70)26. 7025 was converted to a Mother Bus; see sidelight F.

9. Garfield Heights had a unique numbering method in that the first two digits were the number of seats in the coach.

10. RTA AMG coach 8008 and GM 873 are preserved at Ohio Museum of Transportation in Orrville.

11. The following other Cleveland area buses are also preserved at Ohio Museum of Transportation: City of Euclid 38 (REN 1338),

Maple Heights 470 and 1305, and North Olmsted 150. For the celebration of its sixtieth anniversary, employees of the North Olmsted system (NOMBL) restored a 1951 GM bus, number 110, to its original appearance. It is retained by NOMBL.

12. After retirement, a few of the original Argosy CRT buses found a second life as RTA work crew buses for the Power and Plant departments. 9100 REN R400, 9103 REN R403, 9104 REN P404, and 9106 and 9109 were converted but not renumbered. 9409 REN R409 and 9410 REN R410. 9511 was involved in an accident after only 500 miles and never again ran.

13. Buses 2, 4, and 5 were purchased used by RTA from the Cleveland-Lorain Highway Coach Co. (CLC).

14. These three-digit-numbered, orange-painted Columbus buses were ACQ used in August 1987 and were REN as follows (the first two digits representing the year BLT): 930 REN 6530, 949 REN 6649, 104 REN 6804, 112 REN 6812, 115 REN 6815, 118 REN 6818, 120 REN 6920, 123 REN 6923, 129 REN 6929, 130 REN 6930, 132 REN 7032, 135 REN 7035, 136 REN 7036, 137 REN 7037, 140 REN 7040, 145 REN 7045, 551 REN 7551, 578 REN 7558.

15. The 8400s were ACQ used from Columbus in 1990 for establishment of a new loop service around the University Circle area called "The Circle," which operated only briefly.

16. These suburban type 40' buses were purchased to make special high-speed runs from RTA's new Park-n-Ride lots to the downtown area. A different paint scheme of red and blue stripes was applied. Had front door only.

17. Coaches 8318 and 8321 were delivered with Cummins engines as a test and resulted in Cummins engines being supplied in Fleets 47, 88, 90, 91, and 93.

18. These buses equipped with Detroit Diesel natural gas engines.

19. Coaches 4146 and 4158 had GM retrofit air-conditioners.

20. Transportation Manufacturing Corp. (T.M.C.) was a Greyhound Bus–owned company that purchased the GM bus business and continued to manufacture the "Rapid Transit Series" (R.T.S.) buses in Roswell, NM.

RTA ROSTER SIDELIGHTS

A. For a brief period in 1946, CTS operated an articulated Twin-built bus as a demonstration. Briefly in 1982 RTA operated a M.A.N. Truck and Bus Corp.–built demonstration articulated bus. Neither demo resulted in Cleveland adopting the use of articulated buses. Other experimental buses operated over the years included a battery-powered "Electrobus" and a "Steam Bus." In July and August 1995, RTA operated a Motor Coach Industries (MCI) model

102DL3 over-the-road type bus in experimental Park-n-Ride service.

B. City of Euclid bus 1004 was a red/white/blue bicentennial-painted old style GM, and Euclid number 1339 was outfitted as a work bus, had water-bumpers, and was placed by RTA into the miscellaneous equipment roster. 1004 was retained by Euclid.

C. Many of the following Euclid buses were used as replacements for the Shaker Rapid during its rehabilitation in 1980–81: 1441, 1442, 1445, 1505–1508, 1510, 1747–1751, 1753, 1754.

D. The RTA special Zoo Bus, operated as route 20C, was number 2105. Another special was the Easter "Bunny Bus," Flxible number 8547. Special training buses were 4220, 4234, 4235.

E. The RTA Christmas buses were former MARTA 881 and 891.

F. "Mother Buses" held spare parts, with the idea of being able to make emergency repairs on the road in the days before mobile repair trucks. Starting in 1982, two buses were transferred into the miscellaneous equipment classification and CON to "Mother Buses": 1420 REN 01420B and operated from Woodhill Station; 1515 REN 01515 and operated from Triskett Station. In 1983 this fleet was enlarged as follows: Bus #1 from Woodhill REN from 5555, replacing 01420B; Bus #2 from Hayden was REN from 5566; Bus #3 from Triskett continued to be 01515; and Bus #4 from Brooklyn REN from 5556. Brecksville Road Transit single-door suburban-type bus 25, a GM new look, became Mother Bus #5 in 5-1986, replacing 5556, which lost its engine. Prior to that 5556 had been CON to equipment administrator Chuck Kocour's equipment office at Reed garage.

G. DFE Buses. In 1989 RTA embarked on an employee empowerment program for the purpose of increasing ridership called "The Drive for Excellence." Two buses, GM 4149 and 4171, were transformed into "information buses" for use by teams of RTA employees to move about the service area distributing literature and promoting public transit use.

H. The last serviceable bus in CTS colors, 1120, was retired in 1989 and renovated in its blue hue for an historic tribute to transit past and was used in the DFE program in parades. To complete its salute to history, RTA retained three other significant vehicles: first is Shaker Rapid car 12, FN Cleveland streetcar 1212, painted in the original CIRR gray/yellow/red paint scheme; second is Shaker Rapid Pullman PCC car 76, retained in its original condition and presently painted RTA white with red pinstripes; and high-platform "Bluebird" rapid transit car 109, returned to its original blue/silver/red 1954 delivery livery with CTS winged logo. This collection—a bus, a streetcar, a Shaker Rapid car, and a CTS Rapid car—make a handsome package for future generations of transit employees and RTA patrons to treasure.

I. In 1968 coach 1100 was equipped with hard gray plastic seats as a test to see if vandalism costs—which were running $60,000 or more a year—could be reduced. CTS claimed that patrons reacted favorably to these seats, and so one year later when CTS ordered its next fleet, the 3000 series, all 100 units were ordered with the plastic seats.

J. Community Responsive Transit (CRT) became a separately budgeted department in August 1982. Effective September 10, 1995, CRT became known as ADA Paratransit with federally mandated service changes implemented to comply with the Americans with Disabilities Act of 1991.

K. The Transbus Program. In July 1973 the Urban Mass Transportation Administration (UMTA) began a program to redesign the transit bus and add significant innovation to the process. The positive result was introduction by GM of its RTS "rapid transit series" body design, of Rohr-Flxible's "Metro" design, and A. M. General's "Betterbus" design. However, because of the large number of experimental and untested ideas installed in Transbus-designed coaches, many of the mechanical features brought more problems than promise. The RTA 8000 fleet did not live up to its "Betterbus" designation and was retired prematurely. The 4100–4256 GM fleet were RTA's first Transbus GMs, and the problems with the air-conditioners are legion. Confidently delivered with AC and sealed windows, the 4100s became known as the RTA "sauna" buses during the hot summers of 1979–80 when AC failed and windows could not be opened, causing much embarrassment for RTA management and board. After much negative press, the board moved to retrofit the buses with openable windows in 1984 at a $361,000 cost. The situation had become so unsettling that when discussions for further bus purchases were conducted in September 1981, it was requested that bids be submitted for both AC and non-AC options. The board decided not to order any more AC buses, which resulted in the purchase of 336 buses without AC between 1982 and 1988. It was not until Ronald J. Tober became general manager that new AC buses were ordered and retrofit AC units were added to many of the older buses.

L. Full Body Advertising. RTA's first "rolling advertisement" was GM bus 2597, which appeared in a bright-red livery with a huge painted arm bursting through the side to advertise "Today's Power Hits" on radio station Power 108. The next three were Flxible buses; 9239 and 9244 were PTD green with a full-length lizard on the side touting radio station WENZ; coach 9149 served twice, first as a promotion of Cavaliers' basketball, showing a player with ball in hand extending the length of the bus and the slogan "You gotta be there," and then in April 1995 as a Sea World advertisement with

"Shamu 1" PTD in blue and with a killer whale across the length of the bus. Many others followed. 9215, PTD white, became a target for Target Stores. 9220, 9227, 9228, and 9247 became red, white, and blue "Revco—A Friend for Life" buses. Bus 9226 was sky blue, with an airplane for Continental Airways, while 9239 had a white backdrop and large hamburger to promote Rally's. Bus 9240 was festooned with downtown buildings on a green background for First Federal of Lakewood, 9229 bore a huge bagel to toast Smucker's Bagel Toppings, and 9239 served for a second time for Bank One.

Blue Bredas. Full-body treatment was also given to Shaker articulated cars 801–805 when adorned for the city's bicentennial as the Trains of Fame, named for various career paths and then painted with the likenesses of famous Clevelanders: 801, "Business/Industry," featured Frederick C. Crawford, William Otis Walker, Garrett A. Morgan, Lillian and Clara Westrop, John D. Rockefeller; 802, "Leadership," featured Florence Allen, Tom L. Johnson, John P. Green, Carl B. Stokes, James A. Garfield; 803, "Immigration," featured Jane Edna Harris Hunter, Simson Thorman, Henry Chisholm, Theodor Kundtz, Moises Maldonado; 804, "Sports," featured Jesse Owens, John Patrick "Johnny" Kilbane, John W. Heisman, Bennie Friedman, Madeline M. Jackson; 805, "Entertainment," featured Adella Hughes, Frank Hruby, Dorothy Dandridge, Langston Hughes, Effie Ellsler. The first car over the Waterfront Line, 807, was pulled by locomotive 033 on 3-28-96 at 12:30 A.M.; the first energized operation was with line car 032 followed by three-car Breda train consisting of 810, 807, and 809 on 5-23-96 at 9:25 P.M.

M. Vans provided by Yellow Cab Company for short-lived Flats Flyer service.

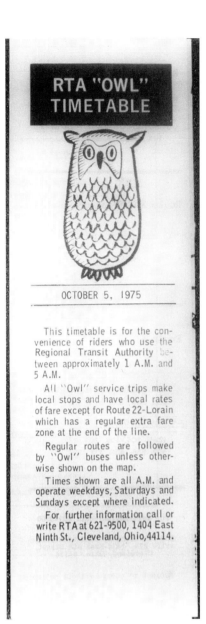

ROSTER SHORT "THE OWL"

All-night service on some main routes has existed almost from the beginning. Cleveland Railway began the practice of publishing a schedule of its "1 A.M. to 5 A.M." service in the 1930s, which came to be known as "owl" service and the vehicle providing the service as the "night car." On today's RTA this service is still provided, but at a premium fare. There is still a printed schedule with a drawing of an owl on the cover, and the vehicle is still called a "night car," even though it is now a bus!

Car 4005 exits the subway at Detroit Avenue and West 29th Street. The car is on a 1951 fan trip. In regular service, Lorain cars used the West 25th Street portal. *(Anthony F. Krisak photo, Richard Krisak Collection)*

CHAPTER THREE

# *The Detroit-Superior Bridge Subway*

JACK AINSLEY

Painting a word picture of what a trip through the old streetcar subway and over the lower deck of the High Level Bridge was like is not a simple job. Such a "picture" would have to project a combination of sights, sounds, and smells. And then, of course, it would change with the time of day or night and with the season.

Entering the subway from any of its three portals was something like entering Cleveland Union Terminal from the east aboard a CTS rapid transit car, but the similarity does not hold with the coming of summer and warm weather. The clatter of steel on steel in the narrow confines of the entrances, particularly on the west side, was almost deafening when car windows were open. And if one happened to be aboard a northbound car, the screech of the wheels on the curve at Detroit was ear shattering. Conversations had to be interrupted entirely whenever two cars passed on the ramp. Even a blind and deaf person would know that he or she was in the West 25th–Detroit subway by feeling its dampness and smelling the burnt electricity.

Put your own imagination to work and picture yourself aboard an eastbound Clifton 1200 grinding along Detroit Avenue at about 4:30 on a summer weekday afternoon. When car 1236 reaches the West 29th Street stop, you observe the motorman opening his destination sign box to change "Clifton–Public Square" to "Clifton-Lakewood," and at almost the same moment a rather bored looking conductor changes his sign, turns on the lights, closes his two doors, and gives a quick two-bell signal before going back to his reverie.

Car 4004 pauses at the West 25th Street subway station. It is August 23, 1951. The sign at the left tells riders that they only have two more days of Detroit Avenue streetcar service. *(Anthony F. Krisak photo, Richard Krisak Collection)*

As the car clatters down the long subway incline just a short distance behind Fulton Road car 1306, you wait for the car to hit the slight bend that brings the tracks into alignment with the High Level Bridge before making your way to the exit for the subway stop. As you get off 1236, your nostrils absorb the damp smell of the place, and your ears are greeted by an almost unceasing roar of wheels.

You are going to transfer to a West 25th streetcar, so you cut over to the passenger tunnel that takes you under the tracks to the southbound platform. Here you wait until your West 25th–Broadview Road car, number 4015, pulling a trailer at this hour, comes along. You board hoping to find a seat, but you are out of luck. The guy next to you echoes your thoughts as you both grab a seat handle to keep from being thrown off your feet as the car makes the ninety-degree turn onto West 25th. "Guess we should have taken the trailer. There are some seats back there."

Your business is completed on the southwest side, so at about 8:00 P.M. you catch a northbound West 25th car, number 4000, at the Pearl

Clark Avenue car 4120 is nearing the West 9th Street subway station. (*Jim Spangler photo*)

and Broadview safety island. You take the first seat on the left-hand side. The smooth, comfortable, gently rocking ride of the freshly painted car is pleasant and rather fast at this hour of the evening. The familiar cross street stops are made with ease: Denison, Trowbridge, Clark, Lorain, and, as the sun is disappearing, Franklin.

The motorman flips the switch that turns on the overhead bullet-type lights and brightly illuminates the interior of the car. Luckily for you he neglects to pull down the shade on his compartment window, leaving you with an unobstructed view of the track ahead. The headlight beam pierces the semi-darkness of the subway as the previously quiet car begins to clatter; the brown cement wall is only a little over a foot from the right-hand windows. Approaching the bottom of the ramp, your motorman cuts his power and gently brakes the rolling steel giant for the turn as blasts of compressed air echo from under the car. At the subway stop two women board, and a sleepy-looking man with a dinner pail gets off. The silent car begins to gently shake to the rhythmic *chunk-a chunk-a chunk-a* sound of the air compressor.

The car picks up power and pulls out onto one of the most fascinating short streetcar rides to be had anywhere in the United States. The round-arched concrete sides of the bridge stand out black against

PCC 4223 is on a "farewell-to-PCCs" fan trip in March 1953. It has just entered the concrete western end of the Detroit-Superior subway. The brighter steel center section is to the rear. *(Bill Vigrass photo)*

*(Opposite)* Clevelanders who can remember the streetcars also remember the big Thanksgiving Day snowstorm of 1950. Those two nostalgic recollections are dramatically illustrated in this snowy scene on Euclid Avenue near East 89th Street. *(Jim Spangler Collection)*

the late evening sky. Just before entering the steel middle section of the bridge, you pass a Madison-Lakewood car, and by turning rapidly you can just catch its number, 4038.

Your eyes have to be busy to catch the downtown skyline, a New York Central streamliner heading for the Terminal, and an ore boat on the dirty river below you. But your attention is now focused on the slow-moving Clifton car two tracks over and going in the same direction. As you catch up with the cumbersome 1200, the two cars are briefly neck and neck, but the trolley drag race is easily won by your fast 4000. Then you glide through an empty West 9th Street station, take the curve, and begin the ascent of the four-track ramp to Superior Avenue, where the emergence of streetlights tells you you've reached busy downtown Cleveland. You've come to the end of a really wonderful streetcar experience.

DETROIT–SUPERIOR BRIDGE SUBWAY · 271

*Cleveland's Transit Vehicles*
was composed in 11½-point Bembo
on a Power Macintosh 7100/80 using PageMaker 5.0
at The Kent State University Press;
printed by sheet-fed offset
on 70-pound Meade Moistrite Matte stock
(an acid-free paper),
Smyth sewn and bound over 98-point binder's boards
covered with ICG Kennett cloth,
and wrapped with dust jackets printed in three colors
on 100-pound enamel stock finished with
matte film lamination
by Thomson-Shore, Inc.;
designed by Will Underwood;
and published by
*The Kent State University Press*
KENT, OHIO 44242